PROTEIN BIOCHEMISTRY, SYNTHESIS,
STRUCTURE AND CELLULAR FUNCTIONS

RECOLLECTIONS OF 45 YEARS IN RESEARCH: FROM PROTEIN CHEMISTRY TO POLYMERIC DRUGS TO THE EPR EFFECT IN CANCER THERAPY

PROTEIN BIOCHEMISTRY, SYNTHESIS, STRUCTURE AND CELLULAR FUNCTIONS

Additional books in this series can be found on Nova's website under the Series tab.

Additional E-books in this series can be found on Nova's website under the E-book tab.

PROTEIN BIOCHEMISTRY, SYNTHESIS,
STRUCTURE AND CELLULAR FUNCTIONS

RECOLLECTIONS OF 45 YEARS IN RESEARCH: FROM PROTEIN CHEMISTRY TO POLYMERIC DRUGS TO THE EPR EFFECT IN CANCER THERAPY

HIROSHI MAEDA

Nova Science Publishers, Inc.
New York

Copyright © 2012 by Nova Science Publishers, Inc.

All rights reserved. No part of this book may be reproduced, stored in a retrieval system or transmitted in any form or by any means: electronic, electrostatic, magnetic, tape, mechanical photocopying, recording or otherwise without the written permission of the Publisher.

For permission to use material from this book please contact us:
Telephone 631-231-7269; Fax 631-231-8175
Web Site: http://www.novapublishers.com

NOTICE TO THE READER

The Publisher has taken reasonable care in the preparation of this book, but makes no expressed or implied warranty of any kind and assumes no responsibility for any errors or omissions. No liability is assumed for incidental or consequential damages in connection with or arising out of information contained in this book. The Publisher shall not be liable for any special, consequential, or exemplary damages resulting, in whole or in part, from the readers' use of, or reliance upon, this material. Any parts of this book based on government reports are so indicated and copyright is claimed for those parts to the extent applicable to compilations of such works.

Independent verification should be sought for any data, advice or recommendations contained in this book. In addition, no responsibility is assumed by the publisher for any injury and/or damage to persons or property arising from any methods, products, instructions, ideas or otherwise contained in this publication.

This publication is designed to provide accurate and authoritative information with regard to the subject matter covered herein. It is sold with the clear understanding that the Publisher is not engaged in rendering legal or any other professional services. If legal or any other expert assistance is required, the services of a competent person should be sought. FROM A DECLARATION OF PARTICIPANTS JOINTLY ADOPTED BY A COMMITTEE OF THE AMERICAN BAR ASSOCIATION AND A COMMITTEE OF PUBLISHERS.

Additional color graphics may be available in the e-book version of this book.

LIBRARY OF CONGRESS CATALOGING-IN-PUBLICATION DATA

ISBN: 978-1-61761-101-8

Published by Nova Science Publishers, Inc. † New York

CONTENTS

Preface		vii
Introduction		1
Chapter 1	Bon Voyage: Passage to the Macromolecular WORLD and Protein Chemistry	3
Chapter 2	Back to Sendai and the World of Antibiotics	5
Chapter 3	On to the United States: Work at Children's Cancer Research Foundation and Harvard Medical School	7
Chapter 4	Kumamoto University in Kyushu, Japan	9
Chapter 5	SMANCS in Clinics and Beyond: The Discovery of the EPR Effect	13
Chapter 6	The Power of Intraarterial Injection of the Polymeric Drug, SMANCS/Lipiodol	15
Chapter 7	Back to the Basics in Infection, Inflammation, and Cancer: The Roles of Proteases, Reactive Oxygen Species (ROS), and Reactive Nitrogen Species (RNS) in Pathogenesis	19
Chapter 8	Clarifying Details of the EPR Effect: A Universal Solid Tumor-Targeting Principle of Macromolecular Anticancer Agents	29
Chapter 9	Factors Facilitating the EPR Effect	35
Chapter 10	Further Enhancement of the EPR Effect	39

Chapter 11	SMA as a Versatile Micelle-Forming Agent	**47**
Chapter 12	Protein Drugs, Then and Now	**51**
Chapter 13	Organizing Various Academic Meetings	**55**
Acknowledgments		**57**
About the Author		**59**
References		**61**
Index		**73**

PREFACE

In this book, I review recollections of my science career that spans almost half a century. During this period, all areas of the biological sciences—including biochemistry, immunology, pharmacology, genetics, cell biology, and oncology—have progressed beyond what one could have imagined 50 years ago, when many of these research areas were almost nonexistent or at an early stage of development.

Biological systems are, in fact, highly complex. As a consequence of multiple interactions within these systems, predictions in such systems are possible only for localized areas or small areas within the ordered complexity. One could, however, follow each evolving step in these systems and be able to deduce a reasonable conclusion or propose a plausible theory. For example, predictions concerning simple, localized in vitro phenomena that is occurring in model systems, are possible in many cases, such as the SCID mouse model, in which a host response such as an immune reaction is minimal if not totally absent. However, predictions of a comprehensive, complex system, such as carcinogenesis in humans, are almost impossible, particularly, which has a nature of unorganized complexity. This idea is similar to those presented by Friedrich August von Hayek in his lecture for the Sveriges Riksbank Prize in Economic Sciences in Memory of Alfred Nobel in 1974[1]. However, I believe that serendipitous findings (predictions) can have more value than findings obtained via rational routes. For students of medicine, therefore, naive ideas are as important as or perhaps better than ideas proposed by an expert with a narrow specialty.

---

[1] From Friedrich August von Hayek—Prize Lecture. Nobelprize.org. 14 Sep 2011. Available at: http://www.nobelprize.org/nobel_prizes/economics/laureates/ 1974/hayek-lecture.html.

The integration of interdisciplinary knowledge or of the interactions of different disciplines becomes more important as medical science expands in various directions. This book demonstrates my understanding of this principle, as it covers my work in diverse research fields, from proteases to free radicals, protein conjugates to vascular permeability, and drug delivery to cancer treatment. Despite their superficial dissimilarity, all these topics are in fact ultimately interconnected.

As I look back over my career, which indeed continues today, I owe a great debt of gratitude to two of my mentors, Robert E. Feeney and Nakao Ishida, both of whom worked in different fields and with whom I spent the very early part of my career, when I was a graduate student. I am, of course, indebted to many more scientists, colleagues, and friends whose advice and suggestions were so valuable for me, my discoveries, and my hypotheses. In addition, although I cannot name here all the students who participated in and supported our research projects, their contributions were key to my achievements.

I must also give full credit to the support and sacrifices of my family, especially my wife Noriko Maeda, ever since I was a Ph.D. student at Tohoku University. Without their encouragement, I might have been much less productive, and much less satisfied, as well. I continually tell students that efforts expended during research are like compound interest: the more you work, the more you gain—not just a linear return but in fact, far greater long-term gains. The generous support of my family made such satisfactory returns possible.

I also had the good fortune to have excellent secretarial support throughout my career, as well as outstanding editorial assistance of Judith B. Gandy, who improved my manuscripts to the highest possible level. Without their assistance, handling the many issues that I faced within and outside the university, as well as my research activities, would not have been possible.

H. M

September, 2011, Kumamoto, Japan

INTRODUCTION

This article describes the lifelong research experience of a scientist, from the beginning, studies of protein chemistry to development of an antitumor protein drug (neocarzinostatin, NCS) to the invention of the first polymer conjugated drug−poly (styrene-co-maleic acid) [SMA] conjugated to NCS, called SMANCS. The author, having acquired knowledge of proteases and inhibitors, pioneered investigations of microbial proteases in the pathogenesis of bacterial infection. The author's group discovered an enormous burst in the generation of superoxide anion radical ($O_2^{·-}$), using a polymer-conjugated enzyme (superoxide dismutase, SOD), during influenza virus infection which triggers the generation of xanthine oxidase. In this setting $O_2^{·-}$ was found to be the major cause of the pathogenesis of this viral infection, which progressed even after the virus was eradiated. These events can be interpreted as advancing the boundary of Robert Koch's postulates, that is, *viral disease occurring in the absence of virus*. Also, the importance of the discovery of endogenous free radicals, now referred to as reactive oxygen species (ROS) and reactive nitrogen species (RNS), during the microbial infections. Role of ROS and RNS in human disease and health became clear, in that they were crucial factors for development of formation of mutant microorganisms (e.g. drug-resistant mutant) in chronic and acute infections as well as carcinogenesis.

In additional studies of the pathogenesis of infections of bacteria and fungi, activation of the kallikrein-kinin cascade and thus bradykinin (also called kinin) generation at the site of infection was found to result in pain and enhancement of vascular permeability (edema). Because no potent inhibitors of bacterial proteases exist in the human body, such proteolytic activity is detrimental to the host. The intense enhancement of vascular permeability of bacterial infection or inflammation was later found to be analogous to the situation in cancer tissues. That finding led to the discovery of the EPR (enhanced permeability and retention) effect of macromolecules (>40 kDa) in solid tumors. This EPR effect can be utilized in the field of cancer chemo therapy that is today known as nanomedicine, for targeting of macromolecular anticancer drugs to tumors. When this type of drug is well developed one could avoid adverse effect and augment the therapeutic efficacy.

The EPR effect is now becoming a universal guiding principle for tumor selective targeting in nanomedicine for design of drugs such as polymer conjugates, liposomes, micelles, antibodies, and DNA/RNA-carrier complexes etc. The polymer conjugate drug SMANCS made remarkable pinpoint targeting possible with unprecedented selectivity: i.e., a tumor/blood ratio of drug concentrations more than 2000 was achieved. Further enhancement, 2- to 3-fold, of the EPR effect, and thereby more tumor selective drug delivery became possible via either angiotensin II-induced elevation of blood pressure or application of nitroglycerin ointment (which produces nitric oxide in tumors). The proof of concept of this therapeutic procedure was demonstrated for advanced cancers of the liver, kidney, bile duct, pancreas, lung in addition to the metastatic cancer; all these cancers are difficult-to-cure type cancer.

As described in the text, such interactions of multiple disciplines and of basic science and clinical problems have yielded many discoveries that will be invaluable to young scientists and future directions and development in the medical sciences.

Chapter 1

BON VOYAGE: PASSAGE TO THE MACROMOLECULAR WORLD AND PROTEIN CHEMISTRY

After obtaining my BS degree from Tohoku University, in Sendai, Japan, in March, 1962, I applied for an extremely competitive Fulbright Graduate Study Fellowship to study in the United States. With great luck, I was accepted into the Fulbright program for study at the University of California, Davis, for 2 years. There, I met Professor Robert E. Feeney, in August, 1962, for the first time; he was to become my mentor throughout my academic career. Professor Feeney, who passed away in September, 2007, was well known for his research on egg white proteins and, more specifically, on ovomucoids and other protease inhibitors. He may be better known, however, for his research on antifreeze glycoprotein found in blood plasma of Arctic and Antarctic fishes. He is also famous in the field of chemical modification of proteins. These research areas in fact became the basis of my own future research directions and developments.

At the University of California, Davis, my MS thesis concerned the molecular microheterogeneity of primarily the egg white protease inhibitors called ovomucoids with genetically identical backgrounds. The remarkable progress in protein chemistry and biochemistry during those days is comparable to that of genomic science at the end of the 20th century. My curiosity about the life sciences led me to choose medical science as my future interest.

Chapter 2

BACK TO SENDAI AND THE WORLD OF ANTIBIOTICS

After finishing my MS degree at Professor Feeney's laboratory, I returned to Sendai, Japan, which is about 350 km north of Tokyo, where I continued my studies at Tohoku University Medical School in Professor Nakao Ishida's Department of Bacteriology. When I had been an undergraduate, in 1960, I had the microbiology class that was taught by Dr. Ishida, who was then an associate professor at Tohoku University Medical School. He later became known throughout the world as the virologist who discovered Sendai virus, who passed away in December 2009 during preparation of this manuscript.

In Sendai, my first research project under Professor Ishida's supervision was to isolate interferon from the allantoic fluid of virus-infected embryonated chicken eggs. At that time, interferon was vaguely known as a protein, but not much more was known. World-class research projects, particularly studies of virus, were conducted in that department; among them, screening program for bioactive compounds with antibacterial, antiviral, or antitumor activity was also actively pursued.

By means of this screening program, Ishida's group discovered a unique antitumor protein, with unprecedented activity, which was named neocarzinostatin (NCS). Because of my former experience in protein chemistry, Professor Ishida assigned me to the investigation of the biochemical and chemical nature of NCS [1]. The biological activity of NCS were studied primarily by Katsuo Kumagai, MD, who had originally been a clinician but then a junior faculty member and later became Professor of Microbiology at Tohoku University School of Dentistry. The Southern Research Institute of the National Cancer Institute in the United States also evaluated the antitumor activity of NCS in the murine Walker 256 tumor model and showed NCS to be one of the most active compounds evaluated among 50 or so new candidate drugs at that time.

To clarify the chemical structure of NCS, we initiated amino acid sequencing of NCS. Use of Sanger's reagent, dinitrofluorobenzene, that is coupled to N-terminal amino acid (to obtain dinitrophenylated derivatives), and Edman's degradation using phenylisothiocyanate was routine procedures. The automatic amino acid analyzer was also becoming a common tool in protein chemistry. In addition, dansyl derivatization with resultant higher fluorescence sensitivity, together with thin layer chromatography, became an accepted method for detection of N-terminal amino acid residues. Despite of inadequate resources, we began the research on amino acid sequencing of NCS. Meantime we developed a highly sensitive amino acid sequencing methods using fluorescein isothiocyanate (FITC), similar to the Edman reagent but much higher sensitivity, for which Hiroshi Kawauchi, a graduate student (a few year junior to me, who later became Professor of Kitazato University and one of the world most expert of the pituitary hormone of fish) contributed a lots, first by synthesizing pure FITC; commercial FITC at that time had a purity of no better than 60%, far from good chemistry. Although we published a few papers, the method did not become popular because no effort for commercialization was made, was one reason [2-4].

Any innovative project or large research undertakings required substantial funding. To obtain such research money, we applied for a research grant to the National Institutes of Health (NIH) of the United States. Several months after submission of the application to the NIH, we received a letter that our application was successful. However, our excitement was temporary. A few months after that letter, we received another letter from the Office of General Accounting stating that the grant could not be funded, perhaps because of the Vietnam War. Student sentiments against the Vietnam War were so strong that they would target the United States-supported research project, even though academic, with violence.

Chapter 3

ON TO THE UNITED STATES: WORK AT CHILDREN'S CANCER RESEARCH FOUNDATION AND HARVARD MEDICAL SCHOOL

Within a few months, Professor Ishida received a letter from Professor Sidney Farber, MD (1903–1973), who was the founder and Director of Children's Cancer Research Foundation (CCRF) of Children's Hospital Boston, at Harvard University Medical School; (CCRF is now Dana-Farber Cancer Institute). The letter stated that he and Dr. Meienhofer's group were interested in NCS and wanted to collaborate in research on NCS. The letter was accompanied by a letter from the late Johannes Meienhofer, PhD, at the same institute, who would be in charge of the project. Such an offer from Harvard was quite appealing, and Dr. Ishida decided that I should go to Boston next year, after my PhD degree was conferred.

Dr. Farber was a pathologist and pediatric oncologist who was well known throughout the world, as Professor at Harvard Medical School, and President of the American Cancer Society (1968–1969). He believed strongly that cancer chemotherapy would be effective, and he developed the concept of total care of pediatric cancer patients, which meant primarily free care and a wide range of support services at CCRF. He pioneered treatments of Wilms' tumor (a kidney cancer) of children with actinomycin D and of lymphoma and leukemia with methotrexate. Dr. Meienhofer was a synthetic peptide chemist working at CCRF, who had been trained in Aachen, Germany, and was involved in the organic synthesis of insulin under Dr. Helmut Zahn.

After I arrived in Boston in May, 1968, I started work on the amino acid sequencing of NCS by using the dansyl-Edman method. To gain more experience in amino acid sequencing, I was sent to the Hormone Research Laboratory of the University of California, San Francisco (UCSF), where Professor Chao Hao Li (1913–1987), who was a friend of Dr. Farber and a scientific consultant to CCRF, was the director. I spent about 2 weeks at the UCSF Hormone Research Laboratory near Golden Gate Park in San Francisco.

What I found was an unusual characteristic of NCS: it resisted chemical reduction of its two disulfide bonds, even under the excess amount of reducing agents, 5 M guanidine-HCl or 8 M urea. To reduce and alkylate NCS, I developed a new method of disulfide bond reduction in liquid ammonia with dithiothreitol, which worked perfectly; this method was published in 1971 in the *Journal of the American Chemical Society* [5]. The remainder of my work involved peptide fragmentation with trypsin, chymotrypsin, and thermolysin, and manual Edman degradation. At that time, high-pressure liquid chromatography for separating peptide fragments was not available, amino acid analysis was not completely automated, and 6 N HCl hydrolysis at 110 °C in vacuo was, of course, required. Tedious labor-intensive efforts were thus prerequisites for amino acid sequencing. About 1 year later, when the project was under way and productive, a postdoctoral fellow, Charles Glaser, joined in our group. When the major portion of the work was almost completed, Dr. Glaser went to UCSF for a second postdoctoral fellowship, and Kenji Kuromizu, another postdoctoral fellow from Kyushu University, Japan, joined our laboratory to help with this project. The total amino acid sequence was completed before long after his arrival [6-8]. The positions of the two disulfide bridges in NCS were determined later, after we returned to Japan [9], and a minor revision of the amino acid sequence [10] was also determined primarily by Dr. Kuromizu.

In the early summer of 1971, I received a letter from Professor Yorio Hinuma, MD, asking me to accept the associate professorship position at his Department of Microbiology at Kumamoto University Medical School. Dr. Hinuma became world known by his discovery of the human T-cell leukemia virus in later year, which is the causative agent of adult T-cell leukemia, for which he received the Hammer Award, among others. He also received the Order of Culture from the Emperor of Japan in 2009. His offer was a challenge for me, inasmuch as the amino acid sequence of NCS was almost completed, so I accepted the offer. I returned to Sendai in the middle of September, to finish work in my previous position, and then in October I moved to Kumamoto to begin work at my new position.

Chapter 4

KUMAMOTO UNIVERSITY IN KYUSHU, JAPAN

I had no acquaintances or relatives in Kumamoto, which is located in the central region of Kyushu Island. Kumamoto University Medical School is an old school with a long history that can be traced back to the 18th century. I continued my studies of NCS, such as chemical modification and structure-activity relationships, biochemical and molecular mechanisms of action, tissue distribution, and pharmacokinetics, pharmacodynamics, and degradation in vivo. One very interesting finding was the highly lymphotropic nature of NCS after subcutaneous injection. When ^{14}C-labeled succinylated NCS was injected subcutaneously, it accumulated primarily in regional lymph nodes [11]. Lymphatic tissue is known to be the preferred site of cancer metastasis, which is why lymphatic metastasis occurs so frequently. This lymphotropic drug accumulation motivated me to devise a strategy for treatment of lymphatic metastasis by utilizing this property: that is, delivering the drug directed to the lymphatic system. In fact, in lymphology, macromolecules and lipids were well known to be preferentially recovered by the lymphatic system after injection into the interstitial space. In other words, lipophilic and macromolecular derivatization of NCS would likely make such lymphotropic drug delivery possible, which would be ideal for an anti-lymphatic metastasis strategy. We also showed that, at a subcellular level, intracellular uptake was needed for DNA degradation by NCS. This finding meant that NCS activity occurred inside cells after entry by endocytosis, rather than signal modulation leading to NCS activity at the surface of the cell membrane [12-14]. Succinylation of NCS suggested that two of the free amino groups in NCS (one at the N-terminal alanine 1 and the other at lysine 20) could be utilized for modification without loss of activity [15,16].

During this work, I found an advertisement in *Chemical and Engineering News* (a weekly journal of American Chemical Society) stating that poly(styrene-co-maleic anhydride) (SMA) could be used as car wax and floor polishing materials, which were water and solvent-soluble, hydrophobic polymers containing maleic anhydride. As was a case of succinic anhydride it would react with amino groups of NCS. Thus, I planned to modify two amino groups of NCS with SMA polymer. I expected that attachment of SMA to NCS would confer a potential (perhaps ideal) lymphotropic property to NCS. For this purpose, I asked ARCO <formerly Atlantic Richfield Chemical Co.> in Philadelphia for a small sample of SMA, and they generously sent me several 1-lb bottles of SMA with different specifications.

Without much difficulty, I reacted SMA (with a relative molecular mass M_r about 6 kDa) and NCS in bicarbonate buffer (pH 8.5), and purified the reaction product, then verified that two SMA chains were conjugated to NCS, which I named SMANCS (Figure 1). The first paper on the synthesis of SMANCS appeared in the *International Journal of Peptide and Protein Research* in 1979 [17]. Descriptions of the biological and pharmacological properties of SMANCS, including its highly lymphotropic nature, were subsequently published [18-20].

Because industrial-grade SMA was unfit for pharmaceutical development, I collaborated with the Kuraray Company in Kurashiki, Japan, which had experts in polymer chemistry and was highly interested in developing pharmaceuticals. After we compared various alkyl side chains of half-alkyl esters of maleic acid residues, we chose to work with the *n*-butyl half-ester of SMA for a number of reasons. The details of this synthesis and the chemistry of SMANCS were published in the *Journal of Medicinal Chemistry* in 1985 [21]. With the fully characterized polymer-conjugated antitumor protein SMANCS in hand, my graduate student Jiro Takeshita (who later became an anesthesiologist) and I were on board for a big voyage on the ocean of macromolecular pharmacology, or polymer therapeutics, as Professor Ruth Duncan later preferred to call it.

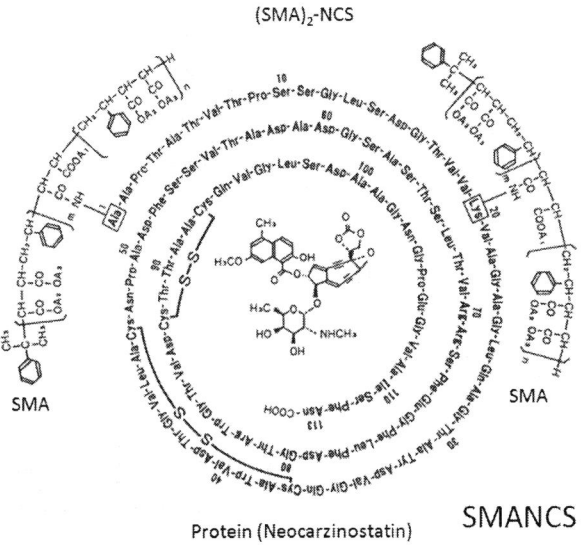

Figure 1. Chemical structure of SMANCS. SMANCS consists of three components: two chains of styrene-co-maleic acid polymer, a protein portion, and the chromophore of NCS.

In the late 1960s, the clinical development of NCS had started a phase I/II study in Japan, although anecdotal accounts of good clinical responses had previously confirmed its efficacy [22,23]. In this regard, my old close friend since college freshman in Tohoku University Medical School, Dr. Ryunosuke Kanamaru, who later became a professor of clinical oncology, was the first doctor to apply NCS in human that is now considered phase 0/I. Professor Nakao Ishida, who was the director of the project, collaborated with Kayaku Antibiotic Laboratories of Tokyo on the manufacturing of NCS and later with Yamanouchi Pharmaceutical Company (now Astellas Pharma Inc.) for marketing. After my return to Japan in 1971, I became a member of the NCS project and was often asked to give lectures on the chemistry, biochemistry, mechanism of action, pharmacology, and pharmacokinetics of NCS. The mode of action of NCS at the molecular level was first studied by my senior colleague, Yasushi Ono, MD, PhD, who demonstrated inhibition of DNA synthesis by NCS, and, at a higher dose, degradation of DNA as its primary mode of action. Dr. Kumagai explored the antitumor effect of NCS in vivo by using various murine models. We confirmed its extremely high potency in bacterial and cultured cell systems: its minimum inhibitory concentration was as low as 0.01 µg · mL^{-1}, or less than 0.1 nM [24,25]. One problem, however, was its very rapid clearance in vivo by both proteolytic degradation and renal excretion.

We later found that this NCS, a small protein of 12 kDa, was being absorbed from the urinary bladder and returned to the blood circulation (by a vesicorenal recirculation mechanism) [26]. Also, high accumulation in the urinary bladder supported its use for bladder cancer. Furthermore, intravenous administration by either a bolus or continuous infusion and by subcutaneous administration led to great differences in plasma drug concentration depending on the routes of administration.

The pharmacokinetics of NCS or proteins in general were still at early stage of pharmacological sciences, and remained to be developed at that time. We later studied its pharmacokinetics and pharmacodynamics in more detail. On the basis of these data, we constructed computer-simulated pharmacokinetic compartment models for slow infusion of NCS via the carotid artery for use in therapy for brain tumors (two-compartment model) and leukemia (one-compartment model) [27,28]. These models were build based on the extremely short half-life in blood and the rapid urinary excretion. Inasmuch as a very low concentration of NCS was infused into the carotid artery for brain cancer, NCS in blood returning to the general circulation would be diluted to subtoxic levels, with concomitant proteolytic degradation help lowering its concentration below nontoxic levels. Later findings also showed that the enhanced permeability and retention (EPR) effect functions in brain cancer, because of the lack of the blood-brain barrier at the tumor site. However, this clinical strategy, based on this unique pharmacokinetic characteristic of NCS, was not fully utilized for brain tumors, except for 19 cases of glioblastoma, for which we obtained good results (unpublished). The reason for this situation was related to economics, as discussed later.

Chapter 5

SMANCS IN CLINICS AND BEYOND: THE DISCOVERY OF THE EPR EFFECT

The drawbacks of NCS, i.e., rapid renal clearance and proteolytic degradation, were overcome by conjugation of NCS with SMA polymer [17-21,29-31]. Also, the antigenicity and immunogenicity against native NCS was almost fully eliminated [20,31] like pegylated proteins. In contrast to the immunosuppression that many anticancer agents are known to cause, SMANCS stimulated natural killer (NK) cells, T cells, and macrophages, and induced interferon owing to its SMA component [32-34]. Newly introduced styrene moiety of SMANCS and SMA also confer the albumin binding property [35], thus albumin masks SMA-conjugates such as SMANCS, and thus they became more biocompatible [35]. SMANCS thus became advantageous to NCS for various reasons.

Although the biological, cytotoxic, and DNA inhibitory activities of SMANCS in vitro were almost equal to those of NCS, in that the molecular size was only 25% larger: 12 kDa versus 16 kDa, SMANCS was markedly different from the parental NCS, particularly in its hydrophobic nature and albumin-binding property. Binding to albumin would cause the M_r of SMANCS to become 90 kDa, so the apparent molecular size of SMANCS in blood plasma would exceed the molecular threshold of renal clearance, and SMANCS would not readily be cleared from the blood like NCS [27,28,31,36]. However, SMANCS leaking out of blood vessels would be cleared via the lymphatic system, as described later, and would thus exhibit lymphotropic behavior.

While we investigated these basic aspects, my colleague, Dr. Toshimitsu Konno of the First Department of Surgery of Kumamoto University Hospital, was trying to treat hepatic and biliary tumors by direct arterial injection via laparotomy and ligation of the hepatic artery. He asked me whether SMANCS would dissolve in Lipiodol, a lipid contrast agent. If it would, SMANCS would be an ideal candidate drug for cancer treatment by arterial injection thereby he would inject SMANCS with Lipiodol. As that time, no anticancer agents were available that were suitable for use with Lipiodol. My answer was yes, it did dissolve in Lipiodol, so we prepared the SMANCS/Lipiodol solution. The next day, Dr. Konno injected this solution into the hepatic artery of rabbits with implanted VX-2 experimental tumors in the liver. When SMANCS/Lipiodol (1.0 mg · mL^{-1}) only 0.1 mL was injected arterially, we recognized its presence in the tumors using a soft X-ray system in the resected liver specimens, even without arterial ligation [37]. Furthermore, no laparotomy was needed, because SMANCS/Lipiodol can be administered by the Seldinger method via the femoral artery. Upon arterial injection via the hepatic artery, SMANCS/Lipiodol could spontaneously penetrated the tumor blood vessels into the interstitial tissue of tumor in a remarkable manner, which was later interpreted as the EPR effect. This finding was quickly applied to clinical settings [38-41].

Chapter 6

THE POWER OF INTRAARTERIAL INJECTION OF THE POLYMERIC DRUG, SMANCS/LIPIODOL

At that time, there was virtually nothing to offer for hepatoma while there were the many hepatoma patients in Japan. One alternative was surgical resection, which would often result in a high and accelerated recurrence of residual tumor except only a few lucky patients. Methods for early diagnosis such as α-fetoprotein (AFP) and computed tomography (CT) were just becoming available late 1970s to early 1980s. Without treatment, patients died, usually within 3–4 months. Because of the poor diagnostic and treatment options, the late Professor Ikuzo Yokoyama, chairman of the Department of Surgery I, asked me whether we could use SMANCS to treat a patient (a 56-year-old woman) by means of Seldinger's method via the femoral artery. With this method, neither laparotomy nor arterial ligation would be necessary, as we found earlier. The patient had an advanced hepatoma, a tumor more than 13 cm in diameter, and her life expectancy, even with blood transfusion and other treatments, was about 2-3 months, but no longer than 4–5 months. After Professor Yokoyama explained this new therapy to the patient and family, we performed this procedure under X-ray guidance, without complications. When the patient appeared in the outpatient clinic for follow-up 3–4 months later, Dr. Konno, who was in charge of the patient, was shocked that the patient was still alive and looked very healthy, and without ascites or jaundice, which were the common manifestations of advanced hepatoma. CT examination showed that the hepatoma had markedly regressed, to an unprecedented extent (Figure 2).

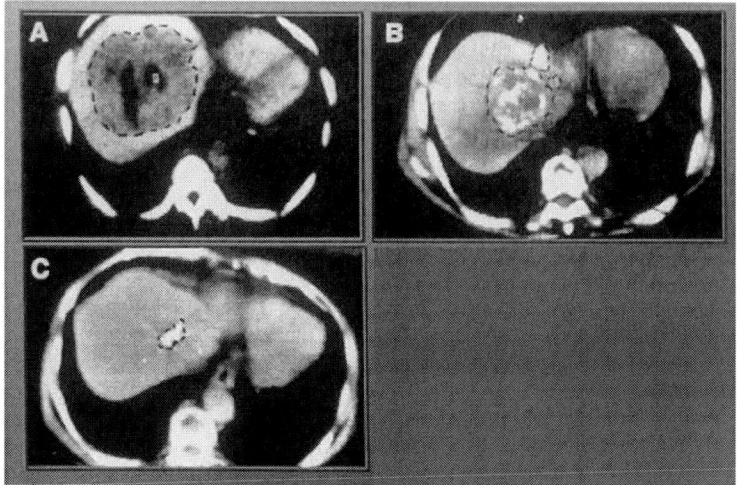

Figure 2. CT images of the liver for the first patient with a hepatoma treated with SMANCS. (A) Before SMANCS application, the dark area encircled by the dashed line is the liver tumor. (B) Four months after SMANCS treatment; (C) 17 months after SMANCS treatment. A marked reduction in tumor size can be seen. Lipiodol appears white high density area indicating tumor area and retaining of SMANCS/Lipiodol.

Dr. Konno rushed into my office with the new and previous CT films, and both of us were of course quite excited. Professor Yokoyama, Dr. Konno, and I were convinced that SMANCS had some magical power. Within 1 year (1982), 22 patients received this treatment, and all but 2 showed a marked reduction in both tumor size and levels of the AFP tumor marker. This result was published in Japanese journal, *Cancer and Chemotherapy* [38], and about 1 year later, a more detailed paper on the clinical effect of SMANCS on hepatocellular carcinoma appeared in an English language journal [39]. These results obviously encouraged all of us involved in development of the drug, and subsequent results were also published [39-41].

The extraordinary efficacy of SMANCS/Lipiodol was related to the truly precise tumor-selective drug delivery, although it required arterial infusion via a designated artery. Along with our excitement about these clinical results, we were interested in a more quantitative evaluation of drug delivery via this technique. Therefore, to analyze the efficacy of this drug delivery, I synthesized a ^{14}C-labeled Lipiodol-moiety from [^{14}C] linoleic acid and administered the ^{14}C-labeled Lipiodol via the hepatic artery targeted to a tumor (VX-2) implanted in the liver of rabbits. Another post graduate student, Ken Iwai, surgeon, MD from Konno's laboratory helped me this part of extensive experiments. Analysis of radioactivity in the tumor confirmed that the highest count—more than 2000-fold—occurred in the tumor compared with that in blood. Furthermore, in contrast to the usual situation for lipid particles in normal tissue, ^{14}C-labeled Lipiodol was not cleared from the tumor tissue as usually occurs via the lymphatic system [36,37]. Indeed, in lymphology, Lipiodol is used to visualize the lymphatic duct and lymph nodes via the X-ray system or by lipid-specific staining. Lipid particles or macromolecules are usually cleared or recovered by the lymphatic system from the tissue interstitium, and the fact that Lipiodol remained in the tumor suggested an impaired lymphatic clearance in the tumor tissues. Furthermore, the distribution of ^{14}C-labeled Lipiodol was only about a few percent of the injected doses in all normal organs such as the colon, kidney, and bone marrow, except for a relatively high value (10–15%) in the spleen and liver which were covered by the hepatic or splenic arteries near the injection route of hepatic artery. This finding indicates that this intraarterial method of SMANCS/Lipiodol administration would almost completely eliminate the systemic adverse effects caused by drug deposition in normal tissues, while producing an extremely effective antitumor effect because the tumor-selective drug accumulation was so marked. The high radioactivity counts of radio labeled lipid in the liver and spleen (normal tissues) were much reduced within a few days, perhaps because of normal lymphatic clearance. Other explanations for the higher radioactivity count in the liver may be the first-pass effect, because the artery, used for the injection, was the one supplying to the liver and the fact that lipid particles are usually recovered via the reticuloendothelial system of the liver and spleen.

Chapter 7

BACK TO THE BASICS IN INFECTION, INFLAMMATION, AND CANCER: THE ROLES OF PROTEASES, REACTIVE OXYGEN SPECIES (ROS), AND REACTIVE NITROGEN SPECIES (RNS) IN PATHOGENESIS

7.A. MICROBIAL DISEASE IN THE ABSENCE OF MICROORGANISMS: MICROBIAL PROTEASES AS PATHOGENS

In the Medical School, I was frequently asked to consult on various clinical problems, namely, important opportunities to interact with members of different clinical departments: Internal Medicine (I/II), Surgery (I/II), Ophthalmology, Dermatology, and Obstetrics-Gynecology, and others including departments in the basic sciences, such as Pharmacology, Anatomy and Pathology. Once, my colleague Professor Ryoichi Okamura, a professor of ophthalmology, showed me severe cases of corneal infection with *Serratia marcescens* and *Pseudomonas aeruginosa*, which did not respond to any antibiotics. The damage to the cornea was so severe and irreversible and the pain was so intense that some patients underwent extraction of the eyeball. Having some knowledge of bacterial proteases, I suggested that bacterial proteases could be the cause. In addition, Dr. Koki Matsumoto and Dr. Ryuji Kamata (who were then graduate students) came to my laboratory to study whether serratia metalloproteinase triggered the kallikrein-kinin cascade, which we found, was initiated by activation of the Hageman factor (or Factor XII) and other factors [42-46]. Human humoral fluid contains no effective inhibitor of any of the proteases of *Serratia* or other bacteria. The end result is pain induced by bradykinin (or kinin) and induction of vascular extravasation of plasma proteins. We were able to measure the extent of vascular leakage of plasma proteins by injecting Evans blue (which bound to albumin in vivo) into guinea pigs; the dye would leak into the interstitial space (outside blood vessels). Then the amount of Evans blue was quantified after extraction from the skin or dermal tissue with formamide. (This method would become an invaluable tool in quantification of vascular leakage from solid tumors during the elucidation of the EPR effect, as described later.) We then found that many microbial proteases activated one or more steps in the kallikrein-kinin cascade, i.e., activation of the Hageman factor, prekallikrein or direct liberation of bradykinin from bradykinin from kininogen. Then, graduate student Akhteruzzaman Molla (who is now Director of the Virus Research Laboratory of Abbott Laboratories, Chicago), from Bangladesh, made significant contributions to this work, after Dr. Matsumoto completed his PhD degree.

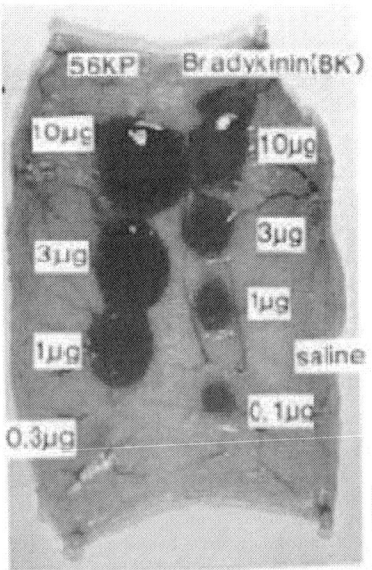

Figure 3. (Continued).

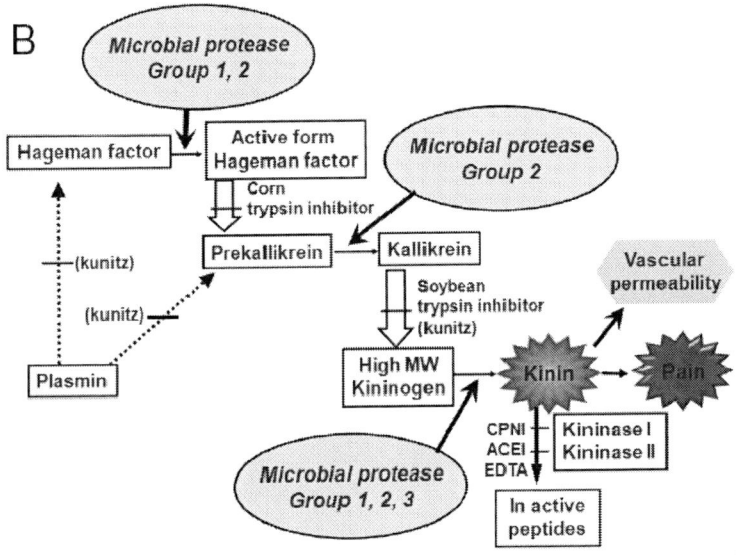

Figure 3. Bradykinin generation by bacterial proteases and activation of kallikrein-kinin cascade. (A) shows extravasation of Evans blue albumin complex induced by intradearmal injection of bradykinin (right) and serratial 56K protease (left). (B) shows kallikrein → kinin cascade and steps that microbial proteases activate. Kunitz, Kunitz type soybean trypsin inhibitor: Corn, corntrypsin inhibitor: CPNI, carboxy peptidase N inhibitor: ACEI, angiotensin conventing enzyme inhibitor: EDTA, ethylenediamine tetra acetic acid.

Extension of our research on bacterial proteases and house dust mite protease produced a surprising and intriguing finding: activation of influenza virus infectivity by cleavage of hemagglutinin on the surface of the virus. For this project, invaluable contributions of two graduate students—Takaaki Akaike and Keishi Maruo—should be mentioned [47,48]. Dr. Akaike (now, Professor of Microbiology, Kumamoto University Medical School) also investigated the true cause of viral pneumonia, as well as the molecular mechanism of double infection with both bacteria and influenza virus, in which proteases caused a more than 100-fold increase in viral infectivity. In our model, we administered, in addition to the virus, bacterial protease, at a dose of about 1 µg, as a bacterial effector. The reproduction of virus was indeed markedly enhanced by proteolytic cleavage of the hemagglutinin of influenza virus, a process that is required for virus infectivity. A surprising finding was that house dust mite protease, which commonly occurs in household air or in the environment of areas inhabited by humans, did increase influenza virus infectivity 100-fold at the 1 µg · mL^{-1} level. This level is a likely concentration of mite protease found in ambient air being inhaled daily, which may be adsorbed on the upper tracheal epithelial surface (per cm^2), the site of influenza virus entry [45-49]. Therefore, in an influenza epidemic, more attention should be paid to exogenous proteases from various bacteria as well as dust derived mite proteases in addition to the endogenous serine-type proteases found in the body (e.g., kallikrein).

This identification of activation of the kallikrein-kinin cascade by bacterial, fungal, and mite-derived proteases at different steps of the cascade was considered an important landmark in the field of kinin study, and it was a major reason for my receiving the Commemorative Gold Medal Award from the E. K. Frey–E. Werle Foundation of Munich, Germany, in 1998. The award was given on the basis of the recommendations of Professor Hans Fritz (Munich) and Professor Werner Müller-Esterl of Mainz (since 2009, President of the Johann Wolfgang Goethe-University Frankfurt am Main) who were experts in this field, and I am most grateful to them [50]. Later, in recognition of the Commemorative Gold Medal Award and my retirement, *Biological Chemistry*, a journal of the German Society for Biochemistry and Molecular Biology produced a special issue (November, 2004). In 1995, the Japanese Society for Bacteriology had presented me with the Asakawa Award, its highest award, for the discovery of the pathogenic roles of bacterial proteases, particularly the finding of the mechanism of kinin generation and the involvement of proteases in the in vivo multiplication of influenza virus.

7.B. MICROBIAL DISEASE IN THE ABSENCE OF MICROORGANISMS: ENDOGENOUS FREE RADICALS AS PATHOGENS

In my department during my chairmanship, we pursued various projects in multiple areas, including bacterial infection and proteases, viral infection and free radicals, and cancer chemotherapy and polymeric drugs. The interaction of different research fields yielded invaluable results, such as the role of polymer-conjugated superoxide dismutase (SOD) for the control of pathogenesis in influenza virus infection. This finding itself then led to the discovery of the superoxide anion radical ($O_2^{\cdot -}$) as a pathogenic molecule. It became the first demonstration of the occurrence of $O_2^{\cdot -}$ in viral disease in the absence of virus [51-53]. This concept went beyond the boundaries of the

Postulates of Robert Koch, which required proof of the presence of defined microbial pathogens at the site of infection or in an ill subject [53, 54].

With reference to a different topic, viral infection triggers a number of events in host defense, or the immune response. One is the oxidative stress that is induced in influenza virus infection, which Dr. Linus Pauling first suggested. However, proving the presence of endogenous free radicals or ROS formation in animals and humans was quite difficult. If one removes ROS, and then as a consequence, one could at least successfully control pathogenesis by using an enzyme that removes superoxide radical; i.e., super oxide dismutase [SOD] which has a molecular size of 30 kDa. In working toward this goal of proving the presence of these radicals, I had enough experience with chemical modifications of NCS, and I knew that the in vivo half-life of small proteins such as SOD would be too short, that native SOD would not work when injected intravenously. Therefore, I designed an SOD to have a longer in vivo half-life by conjugating it with a biocompatible polymer, pyran copolymer [DIVEMA (divinyl ether and maleic anhydride copolymer)]. In collaboration with Dr. Takashi Hirano of Tsukuba, I prepared a pyran-SOD conjugate that had a more than 20-fold longer half-life in plasma after intravenous injection. Dr. Tatsuya Oda, my junior faculty, now Professor of Nagasaki University, and Akaike undertook experiment using influenza mouse model. Injecting this pyran-conjugated SOD into mice infected with influenza virus led to 98% survival, whereas almost all virus-infected control mice died 8–12 days after the infection began (Figure 4A).

It was interesting to follow the time course of viral yield in the lungs of infected control mice: the maximal virus yield was obtained on days 4–5, and from day 8 on mice started to die, but without detectable virus in the lung (Figure 4B). By day 12, all control mice had died, but we found no virus in their lungs (Figure 4A,B). This result means that the cause of viral death and the amount of virus found in the lung were unrelated. In contrast, when we scavenged the $O_2^{\cdot-}$ (or ROS) by means of pyran-SOD, the survival of mice improved greatly, which demonstrated the pathogenic role of $O_2^{\cdot-}$ in influenza virus infection (Figure 4A, B). Then the question was where and how $O_2^{\cdot-}$ generation took place? We then determined that the major source of $O_2^{\cdot-}$ generation was activation of xanthine oxidase in infected lung tissue of the mice (Figure 4C). These results were published in *Science* and other prestigious journal [51-53].

We successfully continued this line of research, and we later found that nitric oxide (NO) is generated in parallel with $O_2^{\cdot-}$ [55] (Figure 4C). Nitric oxide synthase (NOS) generates NO, mostly by the infiltrated leukocytes.

We then realized that $O_2^{\cdot -}$ and NO reacted extremely quickly, at a diffusion-rate manner limited to form peroxynitrite (ONOO$^-$) (Figure 5), which is highly oxidative and acts as a nitrating agent on proteins, nucleic acids, and lipids. We showed for the first time, in viral and bacterial infection models, that the nitration reaction and formation of mutants of virus and bacteria occurred at sites of infection (lesions) [55-57] (Figure 6). This nitration is ubiquitous and affects many aromatic molecules, such as tyrosine (to produce nitrotyrosine) and purines (nitroguanosine), as reported earlier. More interesting and more important, we demonstrated that nitration of guanine at the eighth position led to formation of 8-nitroguanine. 8-Nitroguanosine becomes a substrate of NOSs and cytochrome-P450 reductase etc, and then generate $O_2^{\cdot -}$, thus it lead to propagation reaction yielding ONOO$^-$ and nitration of G, then generation of $O_2^{\cdot -}$ from 8-nitroguanine [57].

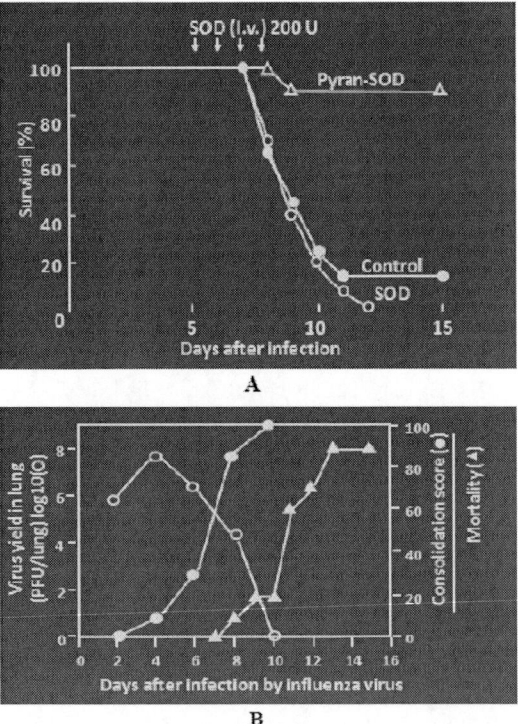

Figure 4. (Continued).

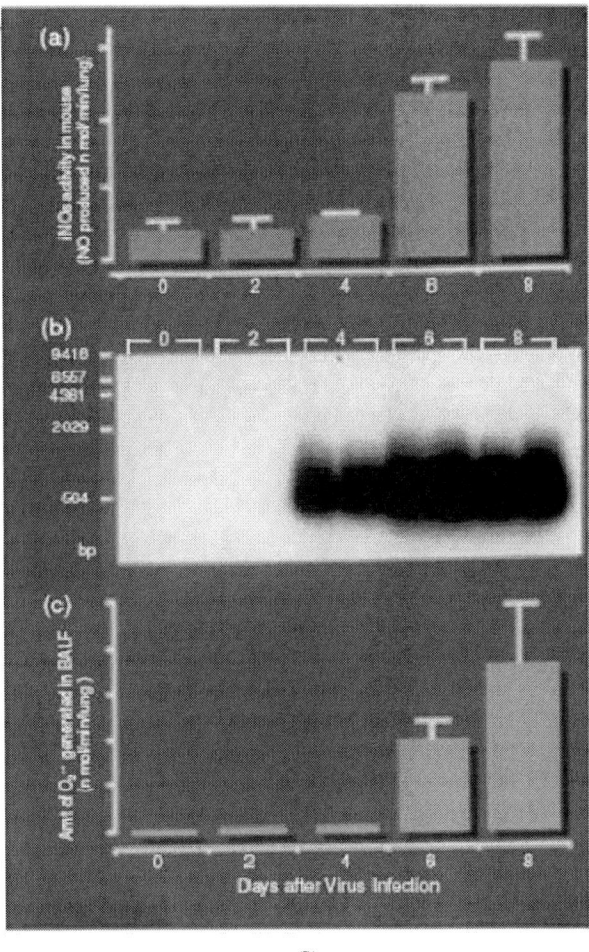

Figure 4. Influenza virus infection in mice. A. Effect of SOD on the survival of infected mice. SOD means mice injected with native SOD. SOD or pyran-conjugated SOD was injected intravenously (ref. [51-53,55]). B. Relationship between virus yield in the lung, consolidation score and mortality. C. (a) NOS activity, (b) amount of NOS in mRNA seen on agarose gel electrophoresis, and (c) generation of super oxide in the infected mouse lungs (see ref. [51-53, 55]).

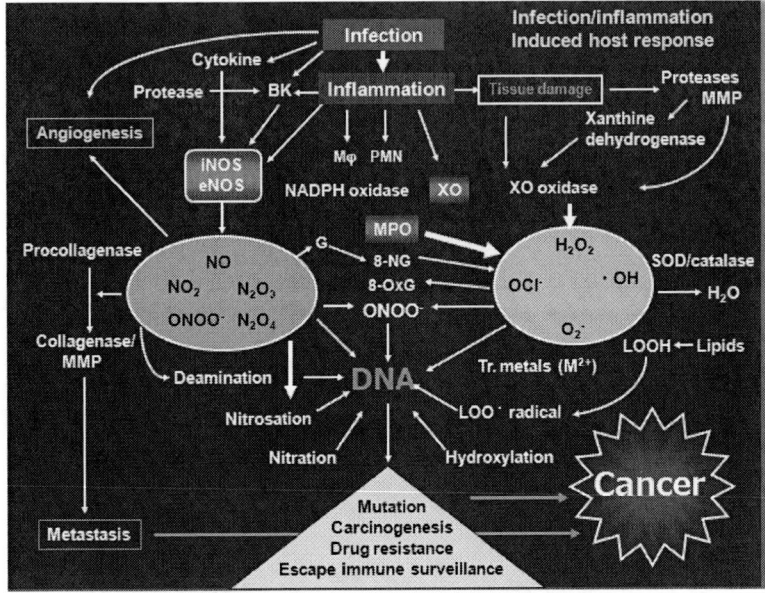

Figure 5. ROS and RNS in infection, inflammation and cancer and their interaction. Formation of ONOO⁻ (peroxynitrite) and damage to DNA, or accelerates mutation and carcinogenesis should be noted.

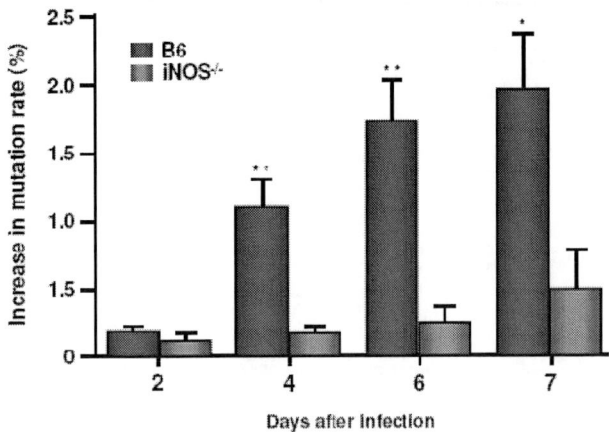

Figure 6. NO dependent increase of mutant virus in GFP-recombinant Sendai virus infected in wild mice (red bar) and not in iNOS knockout mice (green bar).

This result confirmed that the mechanism of mutation occurred at the site of infection, and consequently it also supported inflammation-induced carcinogenesis [54-59]. We reported that exposure to ONOO⁻ caused marked (several-fold) acceleration of viral mutation [57,59-62] (Figure 6), as seen in influenza virus infection [55,56]. Dr. Hideo Kuwahara, a former student in our laboratory, had found ONOO⁻ accelerated mutation in *Salmonella* seen by Ame's test, as well as drug-resistant mutants of *Helicobacter pylori* (at least a 7-fold higher frequency) in the presence of ONOO⁻ at physiological and pathological concentrations [59,60]. Canolol, a phenolic compound that we found in crude rapeseed oil that scavenges ONOO⁻ potently and also know high antioxidant activity, blocked this ROS/RNS-induced mutation [59,60,62], as well as gastric carcinogenesis produced by a carcinogen plus *H. pylori* infection in a *Mongolian gerbil* model, while it suppressed inflammation as revealed by collaboration with Dr. Shoei Tatematsu of Aichi Cancer Center of Nagoya [61]. More recently, Sawa et al reported formation of 3' 5' cyclic-8 nitroguanosine which appears to have a significant role in intracellular signaling pathway [63].

Chapter 8

CLARIFYING DETAILS OF THE EPR EFFECT: A UNIVERSAL SOLID TUMOR-TARGETING PRINCIPLE OF MACROMOLECULAR ANTICANCER AGENTS

At about the same time, in the late 1980s, Dr. Yasuhiro Matsumura, who originally trained as a surgeon and is now at the National Cancer Center Hospital East (head of Drug Development Division), Japan, joined our laboratory to pursue a PhD degree in cancer research. He played a critical role in identifying bradykinin (kinin) —as an effector responsible for facilitating extravasation to form ascetic fluid, in cancer patients (which occurs in carcinomatosis) [64]. This finding stimulated us to investigate the mechanism of such extravasation in cancer tissue in greater detail, because this enhanced vascular permeability was thought to sustain rapid tumor growth by supplying oxygen and nutrients, which may be regarded as the true cause of sustained growth of cancer cells. Moreover, we also thought to utilize this enhanced vascular permeability to control tumor growth, by suppressing kinin generation via either protease inhibitors or kinin antagonists.

My prior knowledge of the chemical modification of proteins was an important element leading to the discovery of the EPR (enhanced permeability and retention) effect [65]. Also, the method of quantitation of extravasation in infectious or inflammatory lesions as described earlier, could be readily applied to investigate extravasation more quantitatively in cancer tissue. Measuring Evans blue in tumor tissue could be viewed as corresponding to measuring delivery of macromolecular drugs of about 67 kDa to tumor tissues, or to accumulation of such drugs in tumor tissues, which we did not see in normal tissues (Figure 7A-(1)). This observation was very intriguing. To validate such preferred accumulation of macromolecules in tumor tissue, we examined the effect of different molecular sizes on tumor uptake of macromolecules in a more detailed manner. The first series of experiments, carried out in mice, involved injecting radio labeled immunoglobulin (IgG, 160 kDa), transferrin (90 kDa), albumin (67 kDa), ovalbumin (47 kDa) and ovomucoid (28.8 kDa) from chicken egg white, and NCS (12 kDa). These substances were primarily labeled with diethylenetriaminepentaacetic acid (DTPA) (a chelating agent), followed by chelation of radioactive ^{51}Cr. The use of this labeling method was described earlier by Professor Claude F. Meares of the University of California, Davis, California. In SMANCS (16 kDa), the two free amino groups of NCS were blocked by conjugation with SMA, so no free amino group was available. Therefore, SMANCS was first cross-linked with free L-lysine via its free carboxyl groups to form amide linkage. DTPA was then added to the newly introduced amino group. With all these radioactive and biocompatible proteins in hand, Matsumura and I undertook extensive experiments. To my relief, my hypothesis that these natural proteins of more than 40 kDa would accumulate more selectively in cancer tissue was confirmed for all large proteins, but not for ovomucoids (< 30 kDa) and NCS, both of which are small proteins. Small proteins were quickly excreted into urine, without uptake by tumors. However, when SMANCS was bound to albumin, its behavior was more like that of true macromolecules, near 90 kDa (Figure 8 A,B).

We described these new findings in a manuscript that we submitted to the journal *Cancer Research* [65]. I believed that this intriguing phenomenon of extravasation and accumulation of macromolecules in tumors and their prolonged residence time in plasma would merit an attractive appellation. I therefore coined the designation *"enhanced permeability and retention effect"* of macromolecules in solid tumor tissues, or the EPR effect, which is now well accepted in the field of drug delivery, particularly in cancer targeting, for examples delivery of liposomes, micelles, antibodies, and DNA/RNA carrier complexes to tumors. The EPR effect is more than just a passive targeting of drugs, however, because its definition includes a prolonged period of drug retention in the tumor tissues. For instance, one can target any cancer drugs or imaging (contrast) agents to solid tumors if one injects them into a tumor-feeding artery; the drugs will be taken up more in tumor tissue but will disappear from the tumor within 5 min or so. This passive targeting is in clear contrast to the EPR effect, because the passive targeting is a temporary phenomenon. In fact, radiologists know this targeting as a tumor stain seen in routine angiography. In our 1986 *Cancer Research* paper [65], we also reported that intratumor retention of Evans blue-albumin injected into tumor directly showed that the dye was far more persistent in tumor than retention in normal tissues. The normal tissues cleared the Evans blue-albumin within a few days or so, and the agent was almost completely gone by 1 week.

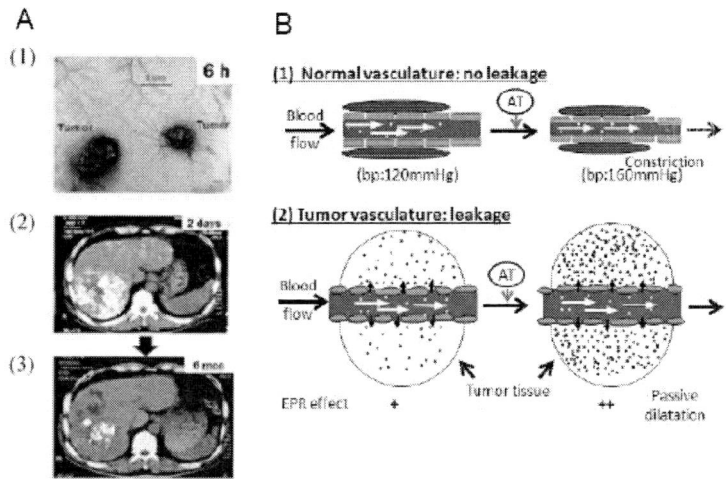

Figure 7. The EPR effect in a tumor in vivo (A) and a schematic illustration of an enhanced EPR effect under angiotensin-II (AT-II)-induced hypertension (B) (reproduced with permission from ref [66]. A: (1) S-180 tumors were implanted in the skin in mice, and when tumor reach to a palpable size Evans blue was injected intravenously. Dark blue spots in (1) demonstrate the tumor-selective accumulation of Evans blue-albumin; this means extravasation of Evans blue-albumin is observed as the EPR effect, which is not seen in normal tissue surrounding the tumors [65]. In Figure 7. A-(2) and -(3), CT scan images of primary hepatoma after SMANCS/Lipiodol injection via the hepatic artery which shows tumor selective drug (SMANCS/Lipidol) uptake by EPR effect in tumor as white area. CT scans were taken 2 days after administration and 6 months after (Figure 7A-(3)). B: The EPR effect does not occur in normal tissue but is observed in cancer tissue [65,66]. Under AT-II-induced hypertension, drug delivery to tumor can be further augmented, with more delivery of the macromolecular drug to tumor, as seen in Figure 7B-(2) on the right, which shows drug being pushed out more [67]. The method is now being applied in clinical settings [66].

The *Cancer Research* paper of 1986 was accepted by the two referees, one of whom wrote a comment directly on the manuscript: "Fantastic findings! Send to the press immediately! ". This paper, authored by Dr. Matsumura and I, was published very quickly, without any revision required. We were, of course, extremely excited. On only a few other occasions in my career did I have papers published in a first-class journal without revision needed. The paper was well received, as was the advent of clinical development of SMANCS. Around this period we were receiving great many hepatoma patients from all over Japan and also from abroad. Maki Clinic in Kikuchi City took care these patients, and both Shojiro Maki, M.D., the director of the clinic and who was also my student, and Dr. Konno were in charge of the clinical procedure.

Cured hapatoma patients from San Antonio, Texas and State of Oklahoma nominated me to become an honorary mayor of the City of San Antonio and an honorary citizen of the State of Oklahoma in 1989. Also, in 1997 the Princess Takamatsu Cancer Research Foundation in Tokyo gave me an award for academic excellence on the basis of development of SMANCS and discovery of the EPR effect in solid tumors.

One great supporter of the EPR effect was Professor Ruth Duncan (at the University of Kiel, then University of London, UK, now at the University of Cardiff). She became an enthusiastic locomotive engine in the field of polymer therapeutics, which is now also called nanomedicine. She and others, including Dr. William Regelson of the Virginia Commonwealth University College of Medicine in Richmond, who had been working on DIVEMA (divinylether maleic anhydride/acid copolymer), also became good friends of mine during this time. Prof. Helmut Ringsdorf at University of Mainz, a renown polymer chemist and was a mentor of Ruth Duncan during her postdoc period, also became a strong supporter of EPR effect. Dr. Duncan and I coorganized a few international symposiums on polymer therapeutics in London and Japan. She sent me a new postdoctoral fellow, Dr. Len Seymour (now Professor at the University of Oxford), to help continue investigations of the EPR effect and other developments in polymer therapeutics. I also received from her well-refined or discrete-sized copolymers of HPMA (hydroxypropyl methacrylate copolymer), which were prepared by the group led by Professor Karel Ulbrich of the Institute of Macromolecular Chemistry, in Prague, Czech Republic. After we radiolabeled these copolymers, we carefully examined their pharmacokinetics and retention in various tissues and tumors with a focus on their molecular size. We at Kumamoto also studied the kinetics of uptake of the copolymers by S-180 tumors, particularly at early time points (i.e., within 1-6 h) [68]; Dr. Duncan's group, then at the University of Birmingham, UK, performed similar studies with B16 melanoma [69]. Other research groups in Tsukuba, Tokyo, and Kyoto have confirmed the EPR effect in different tumor models. Use of HPMA-copolymer of discrete different molecular sizes to analyze the EPR effect thus provided more refined data [36, 68-71].

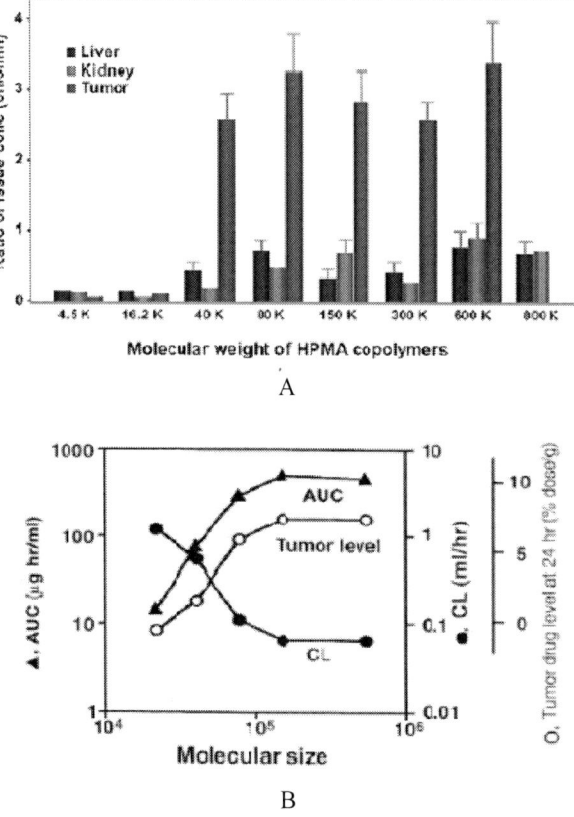

Figure 8. Relationship between the EPR effect and molecular weight of macromolecular drugs. Putative macromolecular drugs used were HPMA copolymers. A. Relative tissue uptake of these polymeric drugs in the liver, kidney and S-180 tumor at 6 hr was compared to that of 6 min after intravenous injection. MW and time dependent increase of tumor uptake is noted [68]. B. Relationship between molecular weight and AUC (area under the concentration curve of blood plasma), tumor uptake and renal clearance (CL). AUC in plasma paralleled to that in tumor [68,71]. A MW larger than 40KDa has a tendency of higher tumor retention.

Another researcher, the late Professor Judah Folkman (1933–2008) at Children's Hospital Boston and Harvard Medical School, also became a strong supporter of the EFR effect. His lifelong study of angiogenesis, tumor neovascularization, and vascular permeability in cancer tissue are integral components of the field of the vascular biology of tumors that he developed. The angiogenesis and EPR effect are the most crucial aspect of tumor growth in view of supplying oxygen and nutrients. I was also most impressed by the excellent scanning electron microscope images of vasculature of tumor blood vessels by Professor Paul O'Brien of the University of Melbourne then (now at Monash University Medical School), who reported leakage of polymers out of tumor blood vessels in *Cancer Research* in 1990 [72] (Figure 9). We still collaborate with his students.

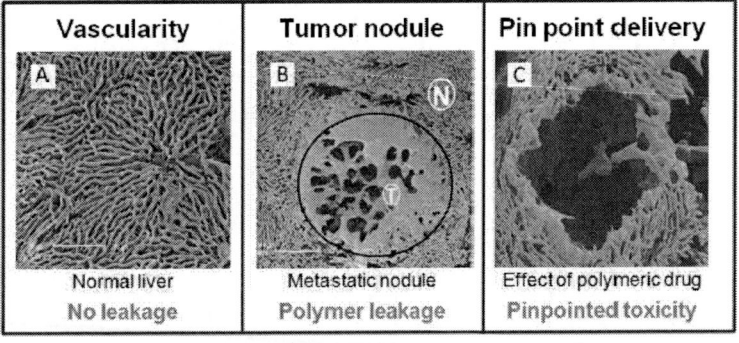

Figure 9. Scanning electron micrographs of vascular casts of plastic resin in the liver. (A) Normal capillary vasucullar structure of the liver. (B) Vascular structure of the liver with a metastatic microtumor nodule, as indicated by o,T. Polymer resin extravasated only into the vascular bed of the tumor, whereas the normal vasculature o,N did not show such polymer leakage. (C) In the same murine tumor model in mouse as that used for B, a macromolecular drug (SMA-pirarubicin micelles) had been injected i.v. 1 week earlier, and that caused selective disintegration of the tumor blood vascular bed (seen as an empty void, equivalent to the circled area containing polymeric resin in the encircled area of tumor in B). The tumor had been chemically induced by dimethylhydrazine in the colon of CBA mice, and a metastasis model was generated by injecting tumor cells into the spleen. Tumor-selective drug delivery and tumor selective damage could be achieved in the mouse even with a tumor size as small as 200 μm, i.e., micronodules. Tumor nodules as small as 200 μm already have unique blood vasculature, which is the evidence of tumor angiogenesis. These pictures are courtesy of Prof. C. Christophi and Ms. J. Daruwalla, University of Melbourne. Parts A and B are adapted with permission from ref. *84,127*.

Chapter 9

FACTORS FACILITATING THE EPR EFFECT

We had previously determined that bacterial infection generated bradykinin via activation of the kallikrein-kinin cascade, which is also a responsible factor for pain and edema formation and frequently accompanies tissue degeneration. The cause of this bradykinin-related pain and edema was identical in both inflammation and cancer. That is, this same mechanism may indeed function in cancer tissue, inasmuch as we had identified excessive bradykinin levels in ascitic and pleural tumor fluids [64,73,74]. These data were important because excessive bradykinin facilitates extravasation of fluid in the cavitary compartment and thus will play a role in ascitic and pleural fluid formation. Dr. Matsumura and Dr. Masami Kimura, who was then also a graduate student, in the Department of Surgery, clarified bradykinin formation in various cancerous ascitic and pleural fluids in many patients [73,74]. Inhibition of bradykinin formation by inhibiting kallikrein with soybean trypsin inhibitor suppressed ascitic fluid accumulation.

On the basis of these findings, I expected that activation of iNOS (the inducible form of NOS) and an accompanying NO generation would occur in tumor tissues. In one of our projects on infectious disease, we had developed a scavenger of NO, i.e., PTIO (2-phenyl-4,4,5,5-tetra methyl imidazoline-1-oxyl-3-oxide) [75]. Using PTIO, we found that NO was also involved in the EPR effect in solid tumors [76-79]. Then, Jun Wu (a graduate student from China, who is now at the City of Hope National Medical Center in California) studied various vascular mediators as well as antagonists and inhibitors of inflammation such as aspirin, indomethacin, kinin antagonists, and L-NMMA (L-N-monomethyl arginine) and confirmed that these mediators affected the EPR effect as well [75-78]. Thus, the EPR effect occurred in cancer as well as in inflammation [73,74-79]. Among these mediators, collagenase and matrix metalloproteinase, activated by ONOO$^-$ (which derived from the reaction of NO with $O_2^{\cdot -}$), were also involved in the EPR effect [78,79]. Other factors that were found to facilitate the vascular permeability of tumors include tumor necrosis factor-α, vascular endothelial cell growth factor (VEGF), which was formally identified as vascular permeability factor by Dvorak and colleagues [80], and interleukin-12. I believe, however, that NO and bradykinin was the most important of all effectors. Therefore, the EPR effect is induced by multiple inflammatory factors (Table 1), and in fact the EPR effect is now demonstrated to function in tumors and inflammatory tissues similarly. Therefore, in terms of permeability and drug-delivery to disease site, various types of macromoleculer drugs or nanomedicines, including antibodies, DNA/RNA carrier complexes, polymer micelles and liposomes, the EPR effect is becoming crucially important, which is seen by the increase of citation numbers of the literature on the EPR effect published in recent ten years (Figure 10) (Table 2). During the preparation of this manuscript (Nov., 2010), we found carbon monoxide (CO) appeared another important factor for facilitating the EPR effect. CO is generated by the heme oxygenase (HO) during the heme catabolism that is carried out by an enzyme HO-1 (also called heat shock protein, Hsp-32), which is highly upregulated in most solod tumors and also in inflammatory state. Thus, HO-1 inhibitor is considered as a good therapeutic target [J. Fang and H. Maeda, in preparation].

Table 1. Factors affecting the EPR-effect of macromolecular drugs in solid tumor

Extensive production of vascular mediators that facilitate extravasation	
a) Bradykinin	b) Nitric oxide (NO)
c) VPF/VEGF	d) Prostaglandins
e) Collagenase (MMPs)	f) Peroxynitrite
g) Carbon monoxide (CO)	h) Inflammatory cells and H_2O_2
i) Tumor necrosis factor (TNF)-α	j) Anticancer agent

Table 2. Architectural differences and functions of tumor vasculature in comparison with normal vasculature

(1) Active angiogenesis and high vascular density

(2) Defective vascular architecture:
• Lack of smooth muscle layer
• Lack of or fewer receptors for angiotensin II
• Large gap in endothelial cell-cell junctions and fenestration
• Anomalous conformation of tumor vasculature (e.g., branching or stretching)

(3) Defective lymphatic clearance of macromolecules and lipids from interstitial tissue (prolonged retention of these substances)

(4) Whimsical and bidirectional blood flow

(5) Macromolecular leakage (> 40 Kda upto size of bacteria) (ref. 65, 68, 71, 81, 82)

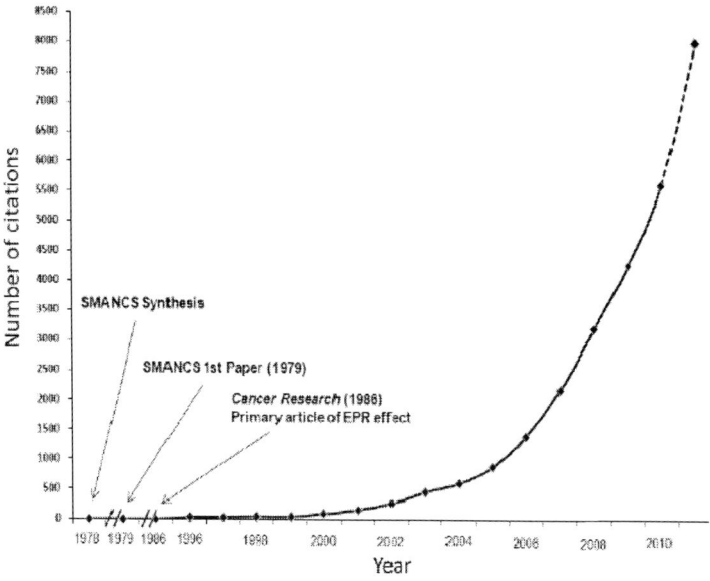

Figure 10. Citation frequency of papers on the EPR effect in solid tumors over time.

The Controlled Release Society awarded me in 2007 the Nagai Innovation Award for Outstanding Achievement on my contribution in the targeted anticancer drug delivery (EPR mechanism) and invention of SMANCS. This award is following after Robert Langer of MIT (2003) and Helmut Ringsdorf (2006). Detailed accounts of EPR effect have reviewed in many occasion [70,71,83-88,118-120]. Japan Society of Drug Delivery System also awarded me Nagai Award in June, 2011.

Chapter 10

FURTHER ENHANCEMENT OF THE EPR EFFECT

The vascular density of rapidly growing tumors such as hepatocellular carcinoma (primary liver cancer) and renal cell carcinoma is usually quite high, whereas that of other tumors, such as pancreatic and prostatic cancer, and metastatic liver cancers, is low. Such hypovascular tumors have lesser vascular density or vascular appearance upon angiography, and they may show a low degree of EPR effect or low vascular density as discussed earlier. People who do not accept the existence of the EPR effect have noted this point as heterogeneity of EPR effect. A hypovascular nature (or heterogeneity of EPR effect) indicates the presence of insufficient angiogenesis for tumor growth. However, blood vessels always occur wherever tissues grow; that is, no tissue lacks blood vessels (with an exception of cartridge). In the window chamber model of solid tumors in rats, one can observe very irregular blood vessels [89], and blood flow may be seen only irregularly, once every, eg. 17 or 21 min unless blood pressure is raised. Hori and colleagues [89,90] also found that blood flow direction may change suddenly. Colloidal osmotic pressure in these solid tumors is believed to become very high and hence suppress penetration of drugs into tumor tissue. One argument proposes that mechanical tissue pressure is generated in an artificial chamber, where tumor cells have a doubling time of 24–48 h, and thus the tissue mass will fill up the chamber and compress the chamber space after a certain time interval, which would thereby impede vascular blood return or normal vascular physiology.

In addition, most tumor cells have adapted hypoxic or anaerobic metabolism for energy production (known as the Warburg effect). Under such conditions, hypoxia-inducible factor, HIF-1α, is known to be activated and to lead to generation of VEGF. Because the EPR effect is the result of vascular leakage of macromolecules from the luminal side of blood vessels to the tumor interstitium to support nutrients and oxygen supply, I initiated further investigations to enhance the EPR effect more by two practical methods for drug delivery to tumors.

10.A. BY ELEVATING BLOOD PRESSURE

The first method involved artificially inducing the hypertensive state (e.g., from 110 to 150 mmHg) with a slow intravenous infusion of angiotensin-II (AT-II). This pathophysiology of tumor vasculature was first described by Prof. Maro Suzuki [90]. Drug in circulation would thus be more effectively pushed into tumor tissue because of the increased vascular pressure would open up only in the tumor vasculature due to incomplete vascular architecture such as lack of smooth muscle layer surrounding the blood vessels (see Figure 7 B(2)).

Regardless of apparently low vascular density tumor, much more blood vasculature can become visible (Figure 11, A→B) when higher blood pressure is generated by slow infusion of angiotensin II (from 100-120 to 150-160 mmHg), by using the angiographic technique (Figure 7 B(2), Figure 11 A, B).

In consistent with this notion, enhanced vascular flow, and hence the drug delivery of SMANCS/Lipiodol to actual human tumor with apparently low vascular tumor (which exhibiting less EPR effect) could be augmented (Figure 11 C,E,G → D,E,F and Figure 12, Case 1-7; Figure 13).

In an AT-II-generated hypertensive state, the endothelial cell-cell gap junctions in normal tissues would become tight, so fewer drugs would leak out of vessels in normal tissues in contrast to tumor vessels which would be opened up (Figure 7B). These experiments were carried out in 1991 by my then-student Chang Li (67), from China, who later received his PhD and MD degrees from Harvard and is now CEO and CSO of Boston Biomedical Inc. and was on the faculty at Harvard Medical School and CEO of ArQule. Clinical evaluation of this method was more recently conducted by my colleagues Dr. Akinori Nagamitsu and Dr. Khaled Greish in our Hakuaikai hospital in Kumamoto, with highly encouraging results that were recently published [66]. Some clinical results of advanced difficult-to-treat tumors are shown in Figure 12 and 13, who were treated under the angiotensin II induced higher blood pressure [66].

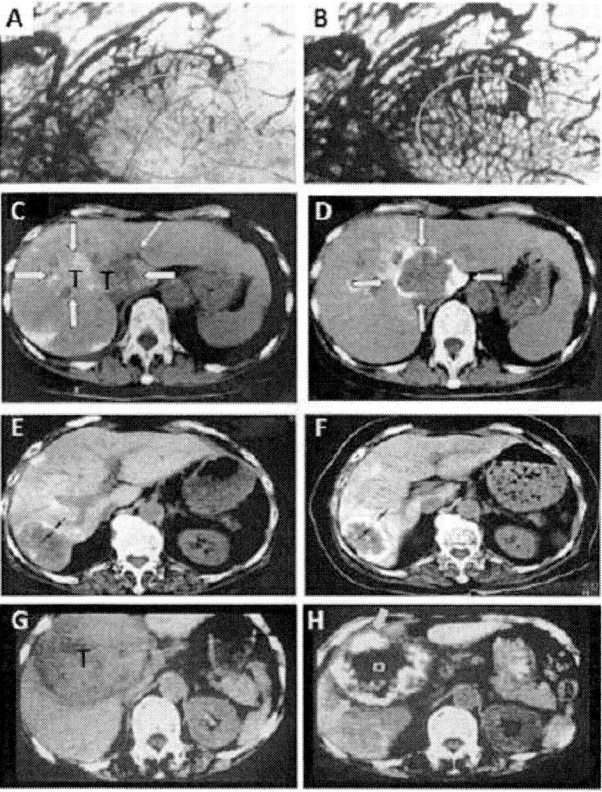

Figure 11. Augmentation of the EPR effect by angiotensin II (AT) induced high blood pressure. Top A/B. A window chamber model of an experimental rat tumor model. Blood vessels are only weekly seen under normotensive state (circled area by pink color) in (A), but the blood vessels became denser as noted in B when the systolic blood pressure of 90 mm was elevated to 160 mmHg (B) (see pink circled area). (adapted from ref. *90,91*). The following examples in Figure 11-13 [66] are results of arterial infusion of SMANCS/Lipiodol® of normotensive blood pressure (90-120 mmHg) (C,E,G) and AT II infused condition (to about 150-160 mmHg)(D,E,F). Both C/D and E/F are colon cancer → liver metastasis; G/H is a case with massive gallbladder cancer metastasized to the liver. In all these (C,E,G) cases are difficult-to-treat cases. Under the angiotensin II induced hypertensive state, CT scan images clearly showed significantly enhanced drug delivery and therapeutic effect [66].

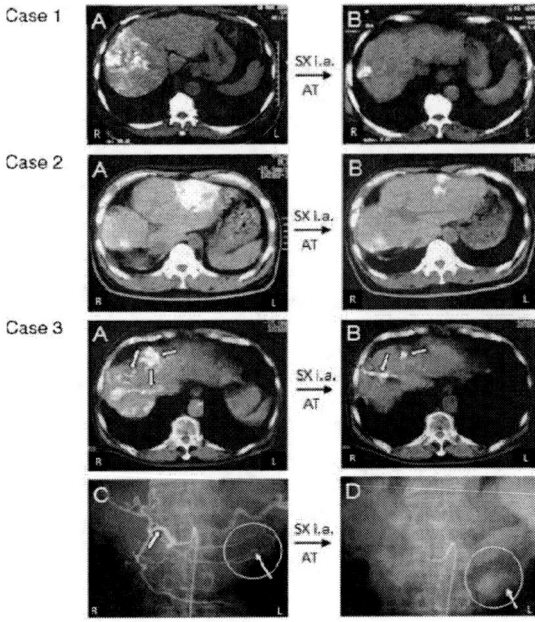

Figure 12. Therapeutic effect of SMANCS (SX)/Lipiodol (LP) on hepatocellular carcinoma (HCC) given via the hepatic artery under the angiotensin II (AT) induced high blood pressure. (Case 1): Response of recurrent HCC to SMANCS/LP under AT-induced hypertension (AT/hyper) 3 years after resection. Reduction in the size of the major tumor in 1 month after only one treatment was remarkable (A vs. B). (Case 2): A massive HCC treated with SMANCS/LP (3 ml) under AT/hyper. A considerable tumor size reduction in 1 month was obtained (A vs. B). (Case 3): Another case of recurrent HCC 2.5 years after resection. The patient received i.a. SMANCS/LP under AT/hyper. The arterio-portal shunts are seen in (A and B). In (C) (angiogram), unique feeding artery branching off from the right proper hepatic artery (arrow) extending to left lower part (encircled tumor) is seen in (D). (B) CT after 6 months, which showed a definite reduction in tumor volume from (A). Enhanced tumor stain (drug delivery) in (D), lower left circled, is seen in the angiogram under AT/hyper administration.

Application of this SMANCS under elevated blood pressure was carried out not only for the primary liver cancer (Figure 12) but also many tumors at advanced stage of metastatic liver cancers from colon cancer, gallbladder cancer (Fig. 11, C, E, G), from stomach cancer, pancreatic cancer, ovarian cancer (Fig. 13, case 4-6), and massive renal cell carcinoma (Fig. 14 (1), (2)) [66]. All cases showed remarkable response [66].

Further Enhancement of the EPR Effect 43

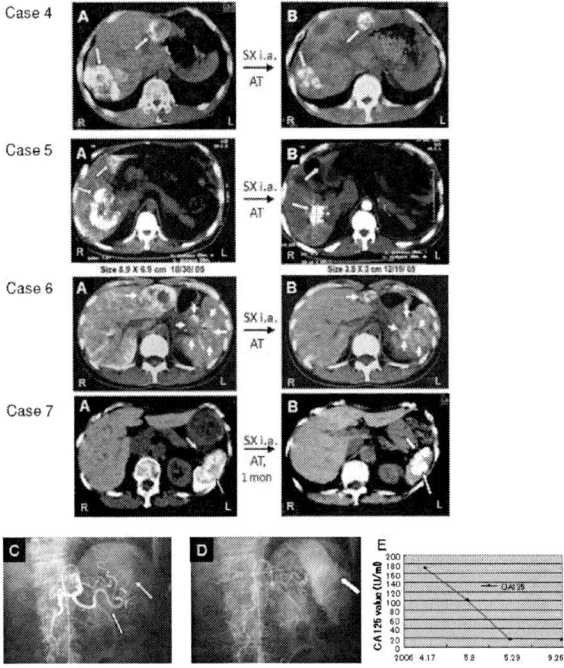

Figure 13. Responses of various metastatic liver cancers to SMANCS/Lipiodol under AT/hyper. (Case 4) A case of stomach cancer metastatic to the liver after resection 3 years earlier. Both posterior and anterior metastatic liver tumors showed marked reductions in tumor volume found 8 months after i.a. SMANCS/Lipiodol infusion under AT/hyper. (Case 5) A massive metastatic liver cancer originating from stomach cancer regressed considerably in 50 days. The B-type staining (peripheral ring shape), usually seen in metastatic tumors by CT, indicates greater drug uptake under AT/hyper (both case 4A and 5A) than under normotensive conditions (data not shown). (Case 6) Massive metastatic liver cancer originating from pancreatic cancer. (A) CT of both the metastatic mass in the liver [center front in (A)] and the primary pancreatic cancer (left middle) at the time of the first infusion of SMANCS/Lipiodol. A large metastatic tumor mass in the frontal area regressed considerably (B) in 5 months after one injection. The primary pancreatic cancer taken up the SMANCS/Lipiodol also (Case 6, Fig. B). (Case 7) Ovarian cancer, of which original tumor had been removed surgically 1.5 yr before. However, it metastasized to the spleen and SMANCS/Lipiodol was infused under AT/hyper (C/D). After 1 month, it showed a good response (B), which corresponded to the decrease of CA125 tumor marker after SMANCS/ Lipiodol (E). Angiography under AT/hyper shows the splenic artery and tumor in the spleen (C and D, white arrow) that is visible in the initial CT scan (A) and significant reduction after 1 month (B and D, heavy tumor stain). Clear radiodense area (arrow in B and D) indicates high uptake of SMANCS/Lipiodol.

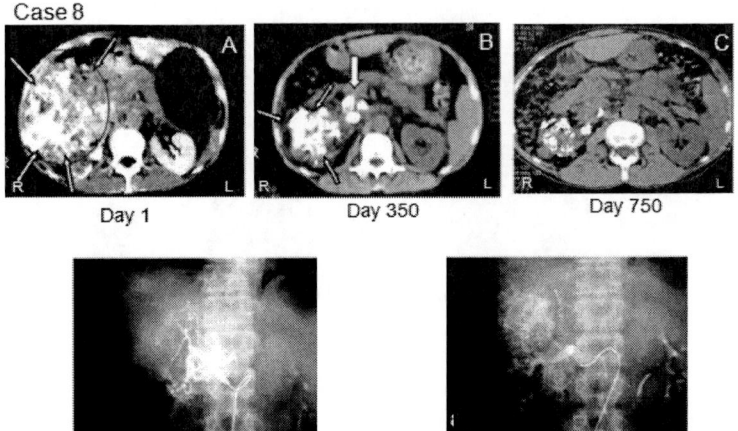

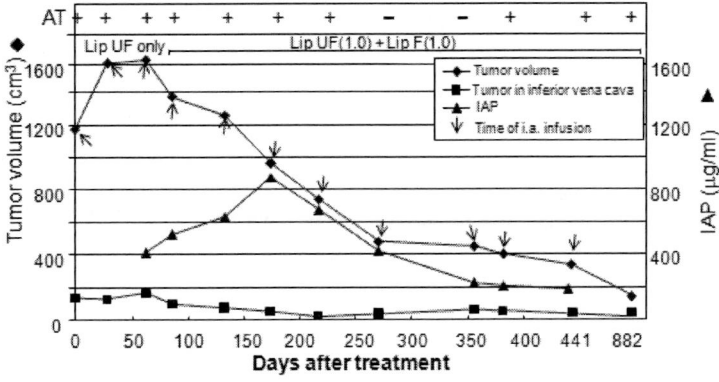

Figure 14. Renal cell carcinoma treated with SX/LP under AT/hyper. (1) (A–C) CT scans of Case 8 on Day 1, Day 350 and Day 750 over time. An initial renal angiogram showing major renal tumor in (D) and also metastatic tumor in the inferior vena cava in (E) seen under angiography, both on Day 1. (2) Time course of tumor volume (closed diamond), tumor marker (immunosuppressive acidic glycoprotein, IAP) values (closed triangle) and estimated volume of metastatic tumor nodule in the inferior vena cava (closed square). The times of i.a. infusions of SMANCS/LP are indicated by short arrows and use of AT/hyper on the top by (+).

Because of my pioneering work on polymeric drugs, particularly SMANCS, and discovery of the EPR effect of macromolecular drugs in solid tumor, I received a Lifetime Achievement Award by *Journal of Drug Targeting*, Informa UK Ltd. (publisher) at the Royal Pharmaceutical Society, in Manchester, in 2007. Also, the *Journal of Drug Targeting* published a special issue to honor my award in 2007. The previous winners of this Award are R. L. Juliano (USA,2004), A. Florence (UK, 2005) and H. Ringsdorf (Germany, 2006).

10.B. BY USING NO RELEASING AGENTS

The second method to enhance EPR effect (or to overcome the heterogeneity of EPR effect) is to utilize NO. We recently found the extremely interesting fact that drug delivery is enhanced by externally applied nitroglycerin (*92,93*). Nitroglycerin has been used for more than 100 years to treat myocardial infarction and angina pectoris. Infarcted myocardial tissue becomes hypoxic, similar to many tumor tissues, so that both tissue oxygen tension (pO_2) and pH values are low if not all cases. Nitroglycerin is readily absorbed from the dermal surface into the circulation, and nitrite ion (NO_2^-) is generated from nitroglycerin (via denitrase) in the hypoxic infarcted tissue. NO_2^- is then converted to NO by nitrite reductase (see Figure 15, below). As described earlier, NO is one of the major vascular permeability factors and facilitates the EPR effect, primarily in tumor tissue. This NO_2^- release thus occurs very similarly or in the same manner, in both infarcted cardiac tissue and tumor tissue. The result is indeed enhanced drug delivery to tumors and an improved therapeutic effect. Takahiro Seki (our postdoctoral fellow), Jun Fang, and I recently reported this finding [92, 93].

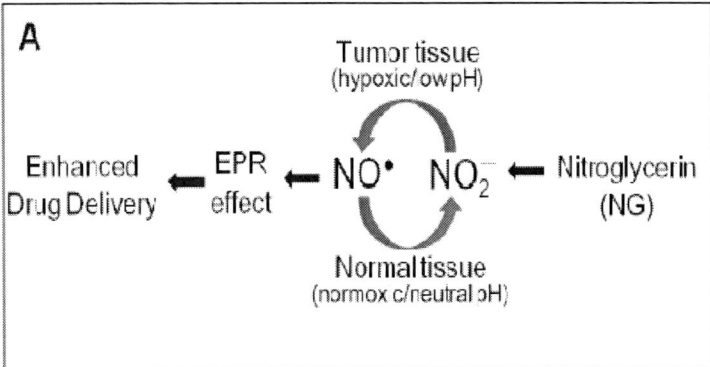

Figure 15. (Continued).

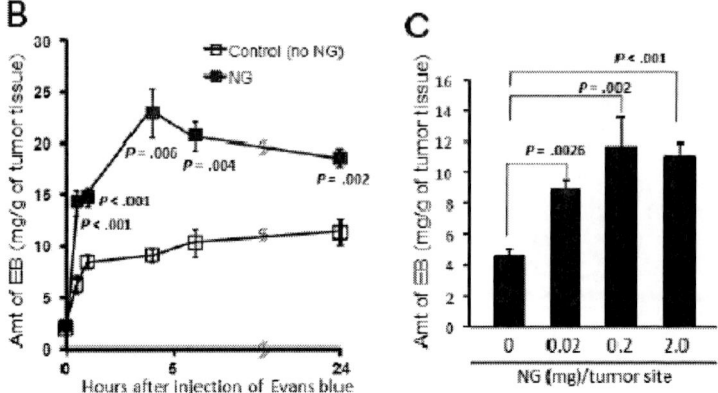

Figure 15. A. Mechanism of nitroglycerin induced augumentation of EPR effect and conversion of nitroglycerin to NO_2^-, then to NO in tumor tissue. B. Increased tumor delivery of Evans blue albumin (macromolecules of 67KDa) in time dependent manner with nitroglycerin (NG) application on the skin. C. Dose dependency of NG in enhancement of EPR effect [93].

Application of this nitrite releasing agent in the clinical setting has been undertaken by Yasuda et al in conventional cancer chemotherapy [94-96]. Their working hypothesis was based on earlier funding that nitroglycerin improved pO_2 of tumor tissue thereby affecting the signaling cascade of cancer tissue including down regulation of vascular endothelial growth factor (VEGF), hypoxia inducible factor and MAP-kinase, etc [97,98]. In different context, since we know NO facilitates EPR-effect, we examined the use of NO releasing agent for enhanced drug delivery and improve clinical effect. We initially infused Nitrol[o,R] (ISDN; isosorbitol dinitrate) to the tumor feeding artery before SMANCS/Lipiodol infusion in human lung cancer patients [99, 101]. The all the results were remarkable. Furthermore, it was found in experimental mouse tumor model system as well as clinical setting, NO alone appears significantly beneficial to suppress tumor growth.

Chapter 11

SMA AS A VERSATILE MICELLE-FORMING AGENT

Dr. Khaled Greish from Egypt joined my department as a graduate student in 2003. He had clinical experience (with a Master's degree in medicine) and a keen interest in new areas of cancer chemotherapy. I assigned him to investigate the possibility of new micelle formation with various anticancer agents using SMA, because of my earlier experience, I knew that SMA formed micelles. The first issue was which drugs to encapsulate in the micelles, and then, how to make the micelles, and what would be the characteristics of the SMA micelles. For a model, we chose the anthracyclines at first: doxorubicin, aclarubicin, taxol, cisplatin and then pirarubicin (tetrahydropyranyl-doxorubicin, or THP). They all formed useful micelles or complex, and particularly with THP, the micelle exhibited slower drug release and more stable than doxorubicin micelles, and indeed, SMA-THP micelles showed excellent in vivo antitumor activity [102-106].

A few years before our investigations of SMA micelles, PEGylated ZnPP (PEG-ZnPP) had been shown to have excellent antitumor properties in vivo; our postdoctoral fellow from India, Dr. S. K. Sahoo [107], and Jun Fang (who is an associate professor, now at Sojo Univ.) in our laboratory described the in vivo evaluation [108]. ZnPP is almost insoluble in water, but PEG-ZnPP and SMA-ZnPP micelles became water-soluble; these ZnPP derivatives exhibit inhibitory effect to heme oxygenase-1 (HO-1). HO-1 is also called heat shock protein (Hsp) 32 and knows as survival factor of cancer cell; HO-1 is highly upregulated in most cancer cells. In these cells, HO-1 generates biliverdin and CO (carbon monoxide) and iron from heme, and biliverdin is then oxidized to bilirubin. We realized that because bilirubin is a potent antioxidant, it confers antioxidative power to tumor cells to protect them against endogenous oxidants (or ROS, reactive oxygen species). ROS may be generated by leukocytes and anticancer agents. Administration of PEG-ZnPP, to suppress bilirubin production in vivo, would thus make target tumor more vulnerable to ROS, thus useful as anticancer agent [107-111].

More recently such encapsulation of zinc protoporphyrin (ZnPP) into SMA- micelles after the work of PEG-ZnPP was found intriguing [104,111-113]. Details such as drug release, stability, tumor targeting via the EPR effect, both therapeutic effect and toxicity in vivo were clarified by J. Fang, K. Greish, and A. K. Iyer (another graduate student from India). All the micellar drugs were better than the parental free drugs in pharmacokinetics, therapeutic effect and toxicity wise. The SMA-pirarubicin micelle (Figure 16) was the best among them: it had a longer (about 200 times) plasma circulation time and a higher (more than 20 times) tumor accumulation than the parent pirarubicin. In tumor-bearing mice, even at one-fifth of the maximal tolerable dose (i.e., a very low dose), SMA-pirarubicin micelles produced 100% survival at more than 200 days without detectable toxicity in S-180 tumor bearing mouse model together with pathological data using the vascular cast and scanning electron microgram, the paper was published recently in Cancer Science as highlighted paper [106].

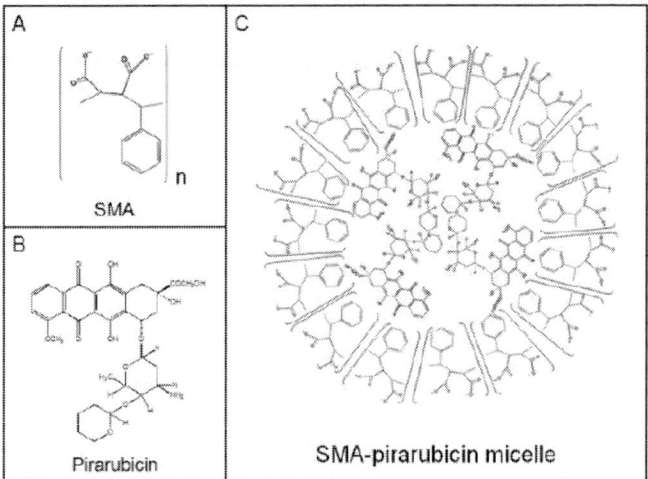

Figure 16. Representation of the chemical structures. (A) Styrene-maleic acid copolymer (SMA). (B) Anticancer agent pirarubicin, THP. (C) Putative structure of SMA-pirarubicin (THP) micelles. Brackets indicate copolymer of SMA (n=10-15).

Our report on the anticancer effects of the HO-1 inhibitor PEG-ZnPP drew the attention of many researchers throughout the world. One of them, Professor Peter Valent (Division of Hematology and Hemostaseology at the Medical University of Vienna), asked me to supply SMA-ZnPP and PEG-ZnPP. Also, from the Humboldt University of Berlin, Professor Beate Raeder (a laser biophysicist) asked me to collaborate. Both collaborations worked very well. Professor Valent's group published a paper on SMA-ZnPP and PEG-ZnPP that showed inhibition of HO-1, and interestingly downregulation of an oncogene *bcl/abl* in chronic myelogenous leukemia cells and other lymphocytic leukemia cells. They showed that SMA-ZnPP micelle was also effective against imatinib-resistant chronic myelogenous leukemia cells. Thus, HO-1 is an ideal molecular therapeutic target for inhibition without toxicity [112-114].

Furthermore, SMA can encapsulate various fluorescent dyes such as fluorescein, rose bengal, and indocyanine green (as recently demonstrated by G. Bharate, a PhD student) as well as chlorophyll studied by Hideaki Nakamura, our junior faculty member) (unpublished data), and we expect future development along this line.

Another important aspect of SMA-micelle we found is that SMA micelles undergo disintegration, and then liberate the drugs from the micelles upon endocytotic intracellular uptake, similar to uncoating of virus capsid/virion upon viral infection; in that process viral genome is liberated free in the cytoplan. This notion was demonstrated experimentally by Hideaki Nakamura recently, which I believe will draw attention of the scientist working on liposome and micelles [115].

We have reported many fluorescent dyes becomes non-fluorescent in SMA micelles due to compact packing in the micelles [102-104] that result in energy transfer and fail to emit fluorescence. Further, the quenched fluorescence can be regenerated by micelle destabilizing agents such as sodium dodecyl sulfate (SDS), ethanol, and lecithin [102, 103, 115]. In the liposome research, it is known that the control of liberation of the encapsulated drugs is one of the most important key point, and people use ultrasonic, heating, or shifting pH, or using environment sensitive polymers, etc to control release velocity. The present finding of drug release upon intracellular uptake is thus so important and ideal. In this setting, amphiphilic component like lecithin in the cell membrane will serve endogenous drug releasing component, therefore SMA micelles, thus appear more ideal than other types of liposomes or micelles.

Chapter 12

PROTEIN DRUGS, THEN AND NOW

During development of proteinaceous pharmaceuticals in medicine, the possible antigenic nature of many proteins has been the first concern, and the labile nature of proteins has been the second. We showed earlier that one could overcome these problems via polymer conjugation, including modification with SMA, which nullified the immunogenicity of SMANCS and prolonged its in vivo half-life [29-31,116-119]. During this period, PEGylation was becoming a standard procedure used to reduce the immunogenicity of proteins. Before NCS was approved in 1971 as a therapeutic agent for treatment of leukemia and cancers of the gastrointestinal tract, pharmaceutical community had very little experience with or knowledge about protein drugs in technical and scientific issues, such as pharmacokinetics, tissue distribution, AUC (area under the concentration-time curve), and inactivation of protein drug.. The protein nature of NCS meant that proteolytic degradation would occur during sample preparation or processing, as well as on administration. Thus, NCS activity would drop rapidly if no protease inhibitor were used [29, 30,116-119]. My experience in Professor Feeney's laboratory helped me clarify these points. However, protein drugs were so new at that time that general acceptance in clinic was not so easy to achieve. NCS has, as mentioned earlier, extremely high activity at or less than the nanomolar range, and is thus very toxic unless the dosage is carefully controlled. Furthermore, its urinary excretion is extremely fast, with its molecular size of about 12 kDa, and its in vivo proteolytic degradation is also very rapid [26,116]. It is thus essential for good therapeutic effect to maintain a meticulously controlled plasma concentration without over shooting the drug level to avoid adverse effect. We published these pharmacokinetic data for the theoretical calculation of infusing velocity for brain tumor via the intracarotid-arterial, and intravenous infusions velocity for leukemia, in which infusion of NCS was determined by considering the rates of elimination (urinary excretion) and inactivation in the blood, and the IC_{50} etc [27,28,121]. However, precise fine-tuning of dosing velocity was not popular among clinicians at that time.

On the occasion of 30 years of the discovery of prototype proteinaceous antitumor agent NCS, we organized a symposium on NCS in Kumamoto in 1994. NCS had become a leading prototype of protein antitumor agents that contains unique endyene chromophore as active moiety that generates reactive oxygen species. Namely, proteinaceous antitumor agents included actinoxanthin in Moscow (Prof. Khoklov), macromomycin in Tokyo (Prof. Umezawa), and lymphomycin and largomycin and others in Sendai (Prof. Ishida). The book on NCS describes the general characteristic of neocarzinostatin, the amino acid sequence and chemical structure of the protein portion and the chromophore, two-dimensional NMR study, X-ray crystallography, the mode of action at the molecular level, and the immunopotentiating effect. Clinical effects and side effects are also discussed, as well as development of the polymer conjugation for the development of anticancer polymer drug SMANCS and its clinical application [121].

Our second macromolecular drug, SMANCS, which the Japanese Government approved for its use in 1993, was launched for marketing in 1994 by Yamanouchi Pharmaceutical Company, Tokyo, but its acceptance was also not so straightforward. One issue was that its approved route of administration was intra-arterial, into the hepatic artery, for treatment of hepatocellular carcinoma (primary liver cancer). A second issue concerned the market size and sales volume. The financial incentive for any drug with a sales volume of less than $10 million or even $100 million dollars (US) per year is not attractive to most pharmaceutical companies, so that not much interest existed for expanding market, or extension for approved therapeutic use for other types of cancers. In addition to these facts, SMANCS required mixing with Lipiodol under ultrasonic at the theater of angiographic procedure as it is supplied separately from Lipiodol, and this made it too awkward step in the busy clinic. Thus, its clinical development other than hepatoma received no support from the manufacture.

Today, a few decades later, protein drugs, many of which are PEGylated proteins, are attracting more attention than ever. According to *Chemical and Engineering News* of July 20, 2009, 4 of the top-selling 15 U.S. pharmaceutical products in 2008 were protein drugs. However, for the category of anticancer drugs of so called molecular target drugs, questions about cost/benefit issues prevail. Namely, many of the molecular target drugs do not satisfy most of the patients. For instance, Tito Fojo and Christine Grady [122], and others [123,124] have expressed concerns about this issue, and The Lancet carried Editorial comment and other article on this matter.

However, I can see great potential for development of anticancer drugs based on the EPR effect, which could target all tumors more selectively and universally. Furthermore, we have now found ways to augment the EPR effect 2- to 3-fold by simple clinical manipulations, as discussed above. Therefore, macromolecular drugs may be more effective than low-molecular-weight drugs, because these EPR effect-based drugs utilize more universal tumor targeting and ubiquitous characteristics of solid tumors as well as inflammatory tissues.

Another issue is related to the system of anticancer drug development, about which one must think more carefully and wisely. For example, the drug-screening mouse models do not truly mimic highly mutated or genetically diversified human solid tumors, because mouse model tumor cell lines have features of primarily one clone and no host reactions (meaning no inflammation → no free radicals → no mutation → no genetic divergence) [125, 126]. I believe that the design of synthetic polymeric drugs will be more economical and effective, which have much better potential of cost/benefit performance than the available contemporary protein drugs such as monoclonal antibodies, which are extremely expensive and selective to one specific target molecule (epitopic) of tumor cells. Thus, these macromolecular drugs will eventually be adopted for use in oncology clinics. In fact, many polymeric drugs are now in phase I and II trials in the United States, Japan, and Europe, and I am excited to see their clinical successes.

Chapter 13

ORGANIZING VARIOUS ACADEMIC MEETINGS

My research activities, in addition to involving bacterial proteases and polymeric drugs, focused on clarifying the roles of endogenous free radicals (e.g., $O_2^{\bullet-}$ and $NO^{\bullet}$) in infection and cancer in the late 1980s to 2004 before the mandatory retirement from Kumamoto University. Incidentally, I organized two international meetings on NO—one in 1997 in Kyoto, Japan, with Professor Noboru Toda (Shiga University) and Professor Salvador Moncada (The Wolfson Institute for Biomedical Research at University College London), and another one, The 3rd International Conference on the Biology, Chemistry, and Therapeutic Applications of Nitric Oxide in 2004 in Nara, Japan, with Professor Mitsuhiro Yokoyama (Kobe University) and Professor Naoyuki Taniguchi (Osaka University). The International Nitric Oxide Society, of which I was president for 2 years, published a special issue of the journal *Nitric Oxide: Biology and Chemistry* to honor my retirement in 2005. This became the third one after *Biological Chemistry* (German Biochemical Society) and *Journal of Drug Targeting* from UK as described. I am also helped to organize four other meetings in Kumamoto, Japan, in the past 7 years: 9th Meeting of the Society of Cancer Prevention of Japan in June, 2002; the 3rd Japan NO Meeting in May, 2003; the 76th Annual Meeting of the Japanese Society for Bacteriology in April, 2003; and the 23rd Meeting of the Japan Society of Drug Delivery System, in June, 2007. Through all these academic activities, I made so many friends nationally and internationally and was a great reward in my scientific carrier.

ACKNOWLEDGMENTS

I am gratefully indebted to the long-term support of The Ministry of Education, Culture, Sports, Science, and Technology, Japan, for my Research Grants, such as Grants-in-Aid for Scientific Research on Cancer Priority Areas including (20015045), Scientific Research (C) (20590049), and others. I am also grateful to all my colleagues, in Kumamoto and elsewhere, Yasuhiro Matsumura, Jiro Takeshita, Toshimitsu Konno, Shojiro Maki, Yosuke Kai, Takaaki Akaike, Koki Matsumoto, Ackter Molla, Keishi Maruo, Masami Kimura, Yoichi Miyamoto, Koichi Doi, Kiyoshi Takahashi, Michio Ogawa, Michio Kawasuji, Ryunosuke Kanamaru, Khaled Greish, Jun Fang, Jun Wu, Jurstine Daruwalla, Christopher Christophi, Ruth Duncan, Judah Folkman, Helmut Ringsdorf, William Regehson, Hans Fritz, Warner Müller-Ester, Karel Ulbrich, Len Seymourwho for the work on proteases and cancer and unnamed in the text for the work on ROS/RNS and infectious diseases; to my wife Noriko; and to Judith B. Gandy, for long-term editing of my manuscripts, and Ms. Asami Takaki for typographical assistance of this manuscript.

ABOUT THE AUTHOR

On November 3, 2010, I was awarded prestigious 70th Nishi-nippon Culture Award from Nishi-nippon Shimbun (News agency in Fukuoka, Japan). On June 27, 2011, The Japan Society of Drug Delivery System awarded me its highest award, The 11th Nagai Award. More recently, on October 5, 2011, Japanese Cancer Association awarded me The Tomozo Yoshida Award of 2011, the most prestigious award of the society for which I am most honored. Each of these award is given to one person a year.

REFERENCES

[1] Maeda, H., Kumagai, K., and Ishida, N. (1966) Characterization of neocarzinostatin. *J. Antibiot. Ser. A 19*, 253-259.

[2] Maeda, H., and Kawauchi, H. (1968) A new method for the determination of N-terminus of peptido chain with fluorescein-isothiocyanate. *Biochem. Biophys. Res. Comm.*, *31*, 188-192

[3] Maeda, H., Ishida, N. Kawauchi, H. and Tuzimura, K. (1969) Reaction of fluorescein-isothiocyanate with protein and amino acids. Part I. *J. Biochem.*, *65*, 777-783

[4] Kawauchi, H., Tuzimura, K., Maeda, H. and Ishida, N. (1969) Reaction of fluorescein-isothiocyanate with protein and amino acids. Part II. *J. Biochem.*, *66*, 783-789

[5] Meienhofer, J., Czombos, J., and Maeda, H. (1971) Reduction of disulfide bonds in liquid ammonia. *J. Am. Chem. Soc.*, *93*, 3080-3081

[6] Meienhofer, J., Maeda, H., Glaser, C. B., Czombos, C., and Kuromizu, K. (1972) Primary structure of neocarzinostatin, an antitumor protein. *Science 178*, 875-876.

[7] Maeda, H., Glaser, C. B., Czombos, J., and Meienhofer, J. (1974) Structure of the antitumor protein neocarzinostatin. Purification, amino acid composition, disulfide reduction, and isolation and composition of tryptic peptides. *Arch. Biochem. Biophys. 164*, 369-378.

[8] Maeda, H., Glaser, C. B., Kuromizu, K., and Meienhofer, J. (1974) Structure of the antitumor protein neocarzinostatin. Amino acid sequence. *Arch. Biochem. Biophys. 164*, 379-385.

[9] Kuromizu, K., Abe, O., and Maeda, H. (1991) Location of the disulfide bonds in the antitumor protein, neocarzinostatin. *Arch. Biochem. Biophys. 286*, 569-573.

[10] Kuromizu, K., Tsunasawa, S., Maeda, H., Abe, O., and Sakiyama, F. (1986) Reexamination of the primary structure of an antitumor protein, neocarzinostatin. *Arch. Biochem. Biophys. 246*, 199-205
[11] Maeda, H., Takeshita, J., and Yamashita, A. (1980) Lymphotropic accumulation of an antitumor antibiotic protein, neocarzinostatin. *Eur. J. Cancer 16*, 723-731.
[12] Maeda, H., Aikawa, S., and Yamashita, A. (1975) Subcellular fate of protein antibiotic neocarzinostatin in culture of a lymphoid cell line from Burkitt's lymphoma. *Cancer Res. 35*, 554-559.
[13] Sakamoto, S., Maeda, H., and Ogata, J. (1979) An uptake of fluorescein isothiocyanate labeled neocarzinostatin into the cancer and normal cells. *Experientia 35*, 1223-1234.
[14] Takeshita, J., Maeda, H., and Koike, K. (1980) Subcellular action of neocarzinostatin: Intracellular incorporation, DNA breakdown and cytotoxicity. *J. Biochem. 88*, 1071-1080.
[15] Maeda, H. (1974) Preparation of succinyl neocarzinostatin. Antimicrob. *Agents Chemoth. 5,* 354-355
[16] Maeda, H. (1974) Chemical and biological characterization of succinyl neocarzinostatin. *J. Antibiotics, 27,* 303-311
[17] Maeda, H., Takeshita, J., and Kanamaru, R. (1979) A lipophilic derivative of neocarzinostatin: A polymer conjugation of an antitumor protein antibiotic. *Int. J. Peptide Protein Res. 14*, 81-87.
[18] Maeda, H., Takeshita, J., Kanamaru, R., Sato, H., Khatoh, J., Sato, H. (1979) Antimetastatic and antitumor activity of a derivative of neocarzinostatin: An organic solvent- and water-soluble polymer-conjugated protein. *Gann 70*, 601-606.
[19] Takeshita, J., Maeda, H., and Kanamaru, R. (1982) In vitro mode of action, pharmacokinetics, and organ specificity of poly(maleic acid-styrene)-conjugated neocarzinostatin, SMANCS. *Gann 73*, 278-284.
[20] Maeda, H., and Matsumura, Y. (1989) Tumoritropic and lymphotropic principles of macromolecular drugs. *Crit. Rev. Ther. Drug Carrier Syst. 6*, 193-210.
[21] Maeda, H., Ueda, M., Morinaga, T., and Matsumoto, T. (1985) Conjugation of poly (styrene-co-maleic acid) derivatives to the antitumor protein neocarzinostatin: Pronounced improvements in pharmacological properties. *J. Med. Chem. 28*, 455-461.
[22] Takahashi, M., Toriyama, K., Maeda, H., Kikuchi, M., Kumagai, K. and Ishida, N. (1969) Clinical trials of a new antitumor polypeptide: Neocarzinostain. *Tohoku J. Exp. Med., 98,* 273-280.

[23] Hiraki, K., Kamimura, O., Takahashi, I., Nagao, T., Kitajima, K., and Irino, S. (1973) Neocarzinostatin, a new chemotherapeutic approach to acute leukemia. *Nouv Rev Fr Hematol 131*, 445-451.
[24] Ono Y., Watanabe, Y., Ishida, N. (1966). Mode of action of neocarzinostatin: inhibition of DNA synthesis and degradation of DNA in *Sarcina lutea*. *Biochim Biophys Acta 119*, 46-58.
[25] Ono, Y., Ito, Y., Maeda, H. and Ishida, N. (1968) Mode of action of neo-carzino-stain-mediated DNA degradation in *Sarcina Lutea*. *Biochim. Biophys. Acta, 155,* 616-618.
[26] Maeda, H., Sakamoto, S., and Ogata, T. (1977) Mechanism of accumulation of the antitumor protein antibiotic neocarzinostatin in bladder tissue: Intravenous administration, urinary excretion, and absorption into bladder tissue. *Antimicrob. Agents Chemother. 11*, 941-945.
[27] Maeda, H., Sano, Y., Takeshita, J., Iwai, Z., Kosaka, H., Marubayashi, T., and Matsukado, Y. (1981) A pharmacokinetic simulation model for chemotherapy of brain tumor with an antitumor protein antibiotic, neocarzinostatin: Theoretical considerations behind a two-compartment model for continuous infusion via an internal carotid artery. *Cancer Chemother. Pharmacol. 5*, 243-249.
[28] Maeda, H., Matsukado, Y., Iwai, Z., Uemura, S., Kuratsu, J., Takeshita, J., Sano, Y. (1982) Pharmacokinetic one-compartment model using neocarzinostatin as a prototype drug and its clinical application to chemotherapy for brain tumor. Part I. Theory and computer simulation for cerebrospinal infusion (in Japanese). *Jpn. J. Cancer Chemother. 9*, 1042-1045.
[29] Maeda, H., and Takeshita, J. (1975) Degradation of neocarzinostatin by blood sera in vitro and its inhibition by diisopropyl fluorophosphate and N-ethylmaleimide. *Gann, 66,* 523-527.
[30] Maeda, H. and Takeshita, J. (1976) Inhibitors of proteolytic enzymes prevent the inactivation by blood of the protein antibiotic neocarzinostatin and its succinyl derivative. *J. Antibiotics, 29,* 111-112.
[31] Maeda, H., Matsumoto, T., Konno, T., Iwaki, K., and Ueda, M. (1984) Tailor-making of protein drugs by polymer conjugation for tumor targeting: A brief review on Smancs. *J. Protein Chem. 3*, 181-193.
[32] Suzuki, F., and Kobayashi, M. (1997) Immunomodulating antitumor mechanism of SMANCS. In *Neocarzinostatin: The Past, Present, and Future of an Anticancer Drug* (eds. Maeda, H., Edo, K., and Ishida, N.,) pp 167-186, Springer, Tokyo.

[33] Masuda, E., and Maeda, H. (1997) Host-mediated antitumor activity induced by neocarzinostatin and its polymer-conjugated derivative in tumor-bearing mice. In *Neocarzinostatin: The Past, Present, and Future of an Anticancer Drug.* (eds. Maeda, H., Edo, K., and Ishida, N.,) pp 187-204, Springer, Tokyo.

[34] Oda, T., Morinaga, T., and Maeda, H. (1986) Stimulation of macrophage by polyanions and its conjugated proteins and effect on cell membrane. *Proc. Soc. Exp. Biol. Med. 181*, 9-17.

[35] Kobayashi, A., Oda. T., and Maeda, H. (1988) Protein binding of macromolecular anticancer agent SMANCS: Characterization of poly(styrene-co-maleic acid) derivatives as an albumin binding ligand. *J. Bioactive Compatible Polymers, 3,* 319-333.

[36] Maeda, H. (2001) SMANCS and polymer-conjugated macromolecular drugs: Advantages in cancer chemotherapy. *Adv. Drug Deliv. Rev. 46,* 169-185.

[37] Iwai, K., Maeda, H., and Konno, T. (1984) Use of oily contrast medium for selective drug targeting to tumor: Enhanced therapeutic effect and X-ray image. *Cancer Res. 44,* 2115-2121.

[38] Konno, T., Maeda, H., Yokoyama, I. *et al* (1982) Use of a lipid lymphographic agent, lipiodol, as a carrier of high molecular weight antitumor agent, SMANCS, for hepatocellular carcinoma. *Cancer and Chemotherapy* 9, 2005-2015 (in Japanese).

[39] Konno, T., Maeda, H., Iwai, K., *et al* (1983) Effect of arterial administration of high-molecular-weight anticancer agent SMANCS with lipid lymphographic agent on hepatoma: A preliminary report. *Eur. J. Cancer Clin. Oncol. 19,* 1053-1065.

[40] Konno, T., Maeda, H., Iwai, K., *et al* (1984) Selective targeting of anti-cancer drug and simultaneous image enhancement in solid tumors by arterially administered lipid contrast medium. *Cancer 54,* 2367-2374.

[41] Maki, S., Konno, T., and Maeda, H. (1985) Image enhancement in computerized tomography for sensitive diagnosis of liver cancer and semiquantitation of tumor selective drug targeting with oily contrast medium. *Cancer 56,* 751-757.

[42] Matsumoto, K., Yamamoto, T., Kamata, R., and Maeda, H. (1984) Pathogenesis of serratial infection: Activation of the Hageman factor-prekallikrein cascade by serratial protease. *J. Biochem. 96,* 739-749.

[43] Kamata, R., Yamamoto, T., Matsumoto, K., and Maeda, H. (1985) A serratial protease causes vascular permeability reaction by activation

of the Hageman factor-dependent pathway in guinea pigs. *Infect. Immun.* 48, 747-753.

[44] Molla, A., Yamamoto, T., Akaike, T, Miyoshi, S., and Maeda, H. (1989) Activation of Hageman factor and prekallikrein and generation of kinin by various microbial proteinases. *J. Biol. Chem.* 264, 10589-10594.

[45] Maeda, H. (1996) Role of microbial proteases in pathogenesis. *Microbiol. Immunol.* 40, 685-699.

[46] Maeda, H. (2002) Microbial proteinases and pathogenesis of infection. *Wiley Encyclopedia of Molecular Medicine* (Creighton, T. E., Ed.) pp 2663-2668, Volume 4, John Wiley & Sons, New York.

[47] Akaike, T., Molla, A., Ando, M., Araki, S., and Maeda, H. (1989) Molecular mechanism of complex infection by bacteria and virus analyzed by a model using serratial protease and influenza virus in mice. *J. Virol.* 63, 2252-2259.

[48] Maruo, K., Akaike, T., Inada, Y., Ohkubo, I., Ono, T., and Maeda, H. (1993) Effect of microbial and mite proteases on low and high molecular weight kininogens. *J. Biol. Chem.* 268, 17711-17715.

[49] Akaike, T., Maeda, H., Maruo, K., Sakata, Y., and Sato, K. (1994) Potentiation of infectivity and pathogenesis of influenza A virus by a house dust mite protease. *J. Infect. Dis.* 170, 1023-1026.

[50] Frizs, H., and Travis, J. (2004) Hiroshi Maeda—40 years of research. Reflections on the occasion of his retirement and 65th birthday. *Biol. Chem.* 385, 987-988.

[51] Oda, T., Akaike, T., Hamamoto, T., Suzuki, F., Hirano, T., and Maeda, H. (1989) Oxygen radicals in influenza-induced pathogenesis and treatment with pyran polymer-conjugated SOD. *Science* 244, 974-976.

[52] Akaike, T., Ando, M., Oda *et al* (1990) Dependence on O_2^- generation by xanthine oxidase of pathogenesis of influenza virus infection in mice. *J. Clin. Invest.* 85, 739-745.

[53] Maeda, H., and Akaike, T. (1991) Oxygen free radicals as pathogenic molecules in viral diseases. *Proc. Soc. Exp. Biol. Med.* 198, 721-727.

[54] Maeda, H. (2000) Paradigm shift in microbial pathogenesis: An alternative to the Koch-Pasteur paradigm in the new millennium. Abstr. for the 13th International Congress of The International Organization for Mycoplasmology, Fukuoka, Japan, 35.

[55] Akaike, T., Noguchi, Y., Ijiri, S. *et al* (1996) Pathogenesis of influenza virus-induced pneumonia: Involvement of both nitric oxide and oxygen radicals. *Proc. Natl. Acad. Sci. U.S.A.* 93, 2448-2453.

[56] Akaike, T., Fujii, S., Kato, A. et al (2000) Viral mutation accelerated by nitric oxide production during infection in vivo. *FASEB J. 14*, 1447-1454.
[57] Sawa, T., Akaike, T., Ichimori, K. et al (2003) Superoxide generation mediated by 8-nitroguanosine, a highly redox-active nucleic acid derivative. *Biochem. Biophys. Res. Comm., 311*, 300-306.
[58] Maeda, H. and Akaike, T. (1998) Nitric oxide and oxygen radicals in infection, inflammation, and cancer. *Biochemistry (Moscow), 63*, (No.7) 1007-1017.
[59] Kuwahara, H., Kariu, T., Fan, J., and Maeda, H. (2009) Generation of drug-resistant mutants of *Helicobacter pylori* in the presence of peroxynitrite, a derivative of nitric oxide, at pathophysiological concentration. *Microbiol. Immunol. 52*, 1-7.
[60] Kuwahara, H., Kanazawa, A., Wakamatsu, D., Morimura, S., Kida, K., Akaike, T., and Maeda, H. (2004) Antioxidative and antimutagenic activities of 4-vinyl-2, 6-dimethoxyphenol (canolol) isolated from canola oil. *J. Agric. Fd. Chem., 52*, 4380-4387
[61] Cao, X., Tsukamoto, T., Seki, T. et al (2008) 4-Vinyl-2,6-dimethoxyphenol (canolol) suppresses oxidative stress and gastric carcinogenesis in *Helicobacter pylori*-infected carcinogen-treated Mongolian gerbils. *Int. J. Cancer 122*, 1445-1454.
[62] J. Yoshitake, T. Akaike, T. Akuta, F. Tamura, T. Ogura, H. Esumi and H. Maeda (2004) Nitric oxide as an endogenous mutagen for Sendai virus without antiviral activity. *J. Virol., 78*, 8709-8719.
[63] Sawa, T., Zaki, M., H., Okamoto, T. et al (2007) Protein S-guanylation by the biological signal 8-nitroguanosine 3',5'-cyclic monophosphate. *Nature Chem. Biol.3,* 727-735.
[64] Maeda, H., Matsumura, Y., and Kato, H. (1988) Purification and identification of [hydroxyprolyl^3]bradykinin in ascetic fluid from a patient with gastric cancer. *J. Biol. Chem. 263*, 16051-16054.
[65] Matsumura, Y., and Maeda, H. (1986) A new concept for macromolecular therapeutics in cancer chemotherapy: Mechanism of tumoritropic accumulation of proteins and the antitumor agent smancs. *Cancer Res. 46*, 6387-6392.
[66] Nagamitsu, A., Greish, K., and Maeda, H. (2009) Elevating blood pressure as a strategy to increase tumor targeted delivery of macromolecular drug SMANCS: Cases of advanced solid tumors. *Jpn. J. Clin. Oncol., 39,*756-766.
[67] Li, C. J., Miyamoto, Y., Kojima, Y., and Maeda, H. (1993) Augmentation of tumor delivery of macromolecular drugs with

reduced bone marrow delivery by elevating blood pressure. *Br. J. Cancer 67*, 975-980.
[68] Noguchi, Y., Wu, J., Duncan, R. Strohalm J, Ulbrich K, Akaike T, Maeda H. (1998) Early phase tumor accumulation of macromolecules: A great difference in clearance rate between tumor and normal tissues. *Jpn. J. Cancer Res. 89*, 307-314.
[69] Seymour, L. W., Miyamoto, Y., Maeda, H., Brereton, M., Strohalm, J., Ulbrich, K., and Duncan, R. (1995) Influence of molecular weight on passive tumour accumulation of a soluble macromolecular drug carrier. *Eur. J. Cancer 31*, 766-770.
[70] Maeda, H., Seymour, L. W., and Miyamoto, Y. (1992) Conjugates of anticancer agents and polymers: Advantages of macromolecular therapeutics in vivo. *Bioconj. Chem. 3*, 351-362.
[71] Maeda, H., Wu, J., Sawa, T., Matsumura, Y., and Hori, K. (2000) Tumor vascular permeability and the EPR effect in macromolecular therapeutics. *J. Control. Release 65*, 271-284.
[72] Skinner, S. A., Tutton, P. J. M., and O'Brien, P. E. (1990) Microvascular architecture of experimental colon tumors in the rat. *Cancer Res. 50*, 2411-2417.
[73] Matsumura, Y., Kimura, M., Yamamoto, T., Maeda, H. (1988) Involvement of the kinin-generating cascade in enhanced vascular permeability in tumor tissue. *Jpn. J. Cancer Res. 79*, 1327-1334.
[74] Matsumura, Y., Maruo, K., Kimura, M., Yamamoto, T., Konno, T., and Maeda, H. (1991) Kinin-generating cascade in advanced cancer patients and in vitro study. *Jpn. J. Cancer Res. 82*, 732-741.
[75] Akaike, T., Yoshida, M., Miyamoto, Y., Sato, K., Kohno, M., Sasamoto, K., Miyazaki, K., Ueda, S., and Maeda, H. (1993) Antagonistic action of imidazolineoxyl *N*-oxides against endothelium-derived relaxing factor/·NO through a radical reaction. *Biochemistry, 32*, 827-832.
[76] Maeda, H., Noguchi, Y., Sato, K., Akaike, T. (1994) Enhanced vascular permeability in solid tumor is mediated by nitric oxide and inhibited by both new nitric oxide scavenger and nitric oxide synthase inhibitor. *Jpn. J. Cancer Res. 85*, 331-334.
[77] Maeda, H., Wu, J., Okamoto, T., Maruo, K., and Akaike, T. (1999) Kallikrein-kinin in infection and cancer. *Immunopharmacology, 43*, 115-128.
[78] Wu, J., Akaike, T., and Maeda, H. (1998) Modulation of enhanced vascular permeability in tumors by a bradykinin antagonist, a

cyclooxygenase inhibitor, and a nitric oxide scavenger. *Cancer Res. 58*, 159-165.
[79] Wu, J., Akaike, T., Hayashida, K., Okamoto, T., Okuyama, A., and Maeda, H. (2001) Enhanced vascular permeability in solid tumor involving peroxynitrite and matrix metalloproteinase. *Jpn. J. Cancer Res. 92*, 439-451.
[80] Senger, D. R., Galli, S. J., Dvorak, A. M., Perruzzi, C. A., Harvey, V. S., Dvorak, H. F. (1983) Tumor cells secrete a vascular permeability factor that promotes accumulation of ascites fluid. *Science, 219*, 983-985.
[81] Kimura N, Taniguchi S, Aoki K, Baba T. (1980) Selective localization and growth of *Bifidobacterium bifidum* in mouse tumors following intravenous administration. Cancer Res. *40*. 2060-2068.
[82] Zhao, M., Yang, M., Ma, H., Li, X., Tan X, Li S, Yang Z and Hoffman R M. (2006) Targeted therapy with a *salmonella* typhimurium leucine-arginine auxotroph cures orthotopic human breast tumors in nude mice. *Cancer Res.* 66, 7647-7652.
[83] Maeda, H., Sawa, T. and Konno, T. (2001) Mechanism of tumor-targeted deliverly of macromolecular drugs, including the EPR effect in solid tumor and clinical overview of the prototype polymeric drug SMANCS. *J. Cont. Release, 74*, 47-61.
[84] Maeda, H., Bharate, G. Y., Daruwalla, J. (2009) Polymeric drugs and nanomedicines for efficient tumor targeted drug delivery based on EPR-effect. *Eur. J. Pharm. Biopharm. 71*, 409-419.
[85] Seki, T., Fang, J., Maeda, H. (2009) Tumor targeted macromolecular drug delivery based on the enhanced permeability and retention effect in solid tumor. In *Pharm. Perspective of Cancers Therapeutics* (eds. Y. Lu and R.I. Mahato), pp 94-120, AAPS Springer Press, New York.
[86] Iyer, A.K., Greish, K., Fang, J., Murakami, R. and Maeda, H. (2007) High-loading nanosized micelles of copoly (styrene–maleic acid) – zinc protoporphyrin for targeted delivery of a potent heme oxygenase inhibitor. *Biomaterials, 28*, 1871-1881.
[87] Greish, K., Fang, J., Inuzuka, T., Nagamitsu, A. and Maeda, H. (2003) Macromolecular anticancer therapeutics for effective solid tumor targeting: Advantages and prospects. *Clin. Pharmacokinetics, 42*, 1089-1105.
[88] Maeda, H. (2001) The enhanced permeability and retention (EPR) effect in tumor vasculature: The key role of tumor-selective macromolecular drug targeting. In *Advances in Enzyme Regulation* (eds. Weber, G.), *41*, pp.189-207, Oxford, UK, Elsevier Science Ltd.

[89] Hori, K., Suzuki, M., Tanda, S., Saito, D., Shinozaki, M., and Zhang, Q. H. (1991) Fluctuations in tumor blood flow under normotension and the effect of angiotensin II-induced hypertension. *Jpn, J. Cancer Res. 82*, 1309-1316.

[90] Suzuki, M., Hori, K., Abe I., Saito, S., and Saito, H. (1981) A new approach to cancer chemotherapy: a selective enhancement of tumor blood flow with angiotensin II. *J. Natl. Cancer Inst. 67*, 663-669.

[91] Suzuki, M., Hori, K., Abe, I., Saito, S., Satoh, H. (1984) Functional characterization of the microcirculation in tumor. *Cancer Metastasis Review. 3*, 115-126.

[92] Maeda, H., Seki, T., Fang, J. (2008) Enhanced tumor delivery of macromolecular antitumor-drugs by topical application of nitroglycerin on superficial tumors. Abst. 67th Annual Meeting of the Jpn. Cancer Assoc. No. P-518, p.213.

[93] Seki, T., Fang, J., and Maeda, H. (2009) Enhanced delivery of macromolecular antitumor drugs to tumors by nitroglycerin application. *Cancer Sci. 100*, 2426-2430.

[94] Yasuda, H., Yamaya, M., Nakayama, K. *et al.* (2006) Randomized phase II trial comparing nitroglycerin plus vinorelbine and cisplatin with vinorelbine and cisplatin alone in previously untreated stage IIIB/IV non-small cell lung cancer, *J. Clin. Oncol. 24*, 688-694.

[95] Yasuda, H., Nakayama, K., Watanabe, M. *et al.* (2006) Nitroglycerin treatment may increase response to docetaxel and carboplatin regimen via inhibitions of hypoxia-inducible factor-1 pathway and P-glycoprotein in patients with lung adenocarcinoma, *Clin. Cancer Res. 12*, 6748-6757.

[96] Yasuda, H., Yanagihara, K., Nakayama, K. *et al.* (2009) Therapeutic applications of nitric oxide for malignant tumor in animal models and human studies, in: B. Bonavida (Ed.), Nitric Oxide and Cancer, Springer Science, New York.

[97] Jordan, B.F., Misson, P.D., Demeure, R. *et al.* (2000) Changes in tumor oxygenation/perfusion induced by the NO donor, isosorbide dinitrate, in comparison with carbogen: monitoring by EPR and MRI, *Int. J. Radiat. Oncol. Biol. Phys. 48*, 565-570.

[98] J.B. Mitchell, D.A. Wink, W. DeGraff, et al., Hypoxic mammalian cell radiosensitization by nitric oxide, *Cancer Res. 53* (1993) 5845–5848.

[99] Iyer, A.K., Khaled, G., Fang, J. and Maeda, H. (2006) Exploiting the enhanced permeability and retention effect for tumor targeting. *Drug Discovery Today, 11*, 812-818.
[100] Nagamitsu, A., Inuzuka, T., Kawasuji, M., Maeda, H. (2003) Recent advances in SMANCS/Lipiodol therapy – enhanced targeting and delivery effect using vascular nodulator, *Drug Delivery System* (in Japanese) *18*, 438-447.
[101] Nagamitsu, A., Khaled, G., Maeda, H. (2010) Therapy of unresectable lung cancer by SMANCS/Lipiodol in combination with isosorbite dinictrate (ISDN) given arterially. Proceeding 69th Ann. Meeting of Jpn. Cancer Assoc. p.102, No.206. Sept. 22-24.
[102] Greish, K., Nagamitsu, A., Fang, F., and Maeda, H. (2005) Copoly (styrene-maleic acid)-pirarubicin micelles: High tumor-targeting efficacy with low toxicity. *Bioconj. Chem.16*, 230-236.
[103] Greish, K., Sawa, T., Fang, J., Akaike, T., and Maeda, H. (2004) SMA- doxorubicin, a new polymeric micellar drug for effective anticancer targeting. *J. Controlled Release 97*, 219-230.
[104] Iyer, A. K., Greish, K., Seki, T., Okazaki, S., Fang, J., Takeshita, K., and Maeda, H. (2007) Polymeric micelles of zinc protoporphyrin for tumor targeted delivery based on EPR effect and singlet oxygen generation. *J. Drug Target. 15*, 496-506.
[105] Daruwalla, J., Greish, K., Wilson, C. M., Muralidharan, V., Iyer, A., Maeda, H., Christophi, C. (2009) Styrene maleic acid-pirarubicin disrupts tumor microcirculation and enhances the permeability of colorectal Liver Metastases, *J. Vascu. Res., 46*, 218-228.
[106] Daruwalla, J., Nikfarjam, M., Greish, K. Malcontenti-WilsonI, C., Muralidharan, Christophi, V. C. and Maeda, H. (2010) In vitro and in vivo evaluation of tumor targeting SMA-pirarubicin micelles: Survival improvement and inhibition of liver metastases. *Cancer Science, 101*, 1866-1874.
[107] *Sahoo, S. K., S*awa, T., Fang, J., Tanaka, S., Miyamoto, Y., Akaike, T., and Maeda, H. (2002) Pegylated zinc protoporphyrin: A water-soluble heme oxygenase inhibitor with tumor-targeting capacity. *Bioconj. Chem. 13*, 1031-1038.
[108] Fang, J., Sawa, T., Akaike, T., Akuta, T., Sahoo, S. K., Griesh, K., Hamada, A., Maeda, H. (2003) In vivo antitumor activity of pegylated zinc protoporphyrin: Targeted inhibition of heme oxygenase in solid tumor. *Cancer Res. 63*, 3567-3574.
[109] Doi, K., Akaike, T., Fujii, S., Tanaka, S., Ikebe, N., Beppu, T., Shibahara, S., Ogawa, M. and Maeda, H. (1999) Induction of haem

oxygenase-1 by nitric oxide and ischaemia in experimental solid tumours and implications for tumour growth. *Br. J. Cancer*, 80, 1945-1954.

[110] Fang, J., Sawa,T., Akaike,T., Greish, K. and Maeda, H. (2004) Enhanced chemotherapeutic effect against tumor cells by a heme oxygenase inhibitor, pegylated zinc protoporphyrin in combination with anticancer agents. *Int. J. Cancer 109*, 1-8.

[111] Fang, J., Akaike, T., and Maeda, H. (2004) Antiapoptotic role of heme oxygenase and its potential as an anticancer target. *Apoptosis*, 9, 27-35.

[112] Regehly, M., Greish, K., Rancan, F., Maeda, H., Böhm, F. and Röder, B. (2007) Water-soluble polymer conjugated of ZnPP for photodynamic tumour therapy. *Bioconj. Chem.*, *18*, 494-499.

[113] Mayerhofer, M., Gleixner, K. V., Hörmann, G., Sonneck, K., Aichberger, K. J., Ott, R. G., Greish, K., Nakamura, H., Derdak, S., Samorapoompichit, P., Pickl, W. F., Sexl, V., Esterbauer, H., Sillaber, C., Maeda, H., and Valent, P. (2008) Targeting of heat shock protein 32 (Hsp32)/heme oxygenase-1 (HO-1) in leukemic cells in chronic myeloid leukemia: A novel approach to overcome resistance against imatinib. *Blood* 111, 2200-2210.

[114] Hadzijusufovic, E., Rebuzzi, L., Gleixner, K. V., Ferenc, V., Kondo, R., Gruze, A., Kneidinger, M., Krauth, M.-T., Mayerhofer, M., Samorapoompichit, P., Greish, K., Pickl, W. F., Maeda, H., Willmann, M., and Valent, P. (2008) Targeting of heat-shock protein 32/heme oxygenase-1 in canine mastocytoma cells is associated with reduced growth and induction of apoptosis. *Exp. Hematol. 36*, 1461-1470.

[115] Nakamura, H., Fang, J., Bharate, G., Tsukigawa, K., Maeda, H. Intracellular uptake and behavior of two types zinc protoporphyrin (ZnPP) micelles, SMA-ZnPP and PEG-ZnPP as anticancer agents; Unique intracellular disintegration of SMA micelles (in press, 2011).

[116] Maeda, H., Yamamoto, N. and Yamashita, A. (1976) Fate and distribution of [^{14}C] succinyl neocarzinostatin in *rats. Europ. J. C*ancer, 12, 865-870

[117] Maeda, H., Kimura, M., Sasaki, I., Hirose, Y. and Konno, T. (1992) Toxicity of bilirubin and detoxification by PEG-bilirubin oxidase conjugate: A new tactic for treatment of jaundice. In *Poly (Ethylene Glycol) Chemistry: Biotechnical and Biomedical Applications* (ed. J. M. Harris) pp.153-169, Plenum Press, New York.

[118] Kimura, M., Matsumura, Y., Konno, T., Miyauchi, Y. and Maeda, H. (1990) Enzymatic removal of bilirubin toxicity by bilirubin oxidase *in*

vitro and excretion of degradation products *in vivo*. *Proc. Soc. Exp. Biol. Med.*, *195*, 64-69.
[119] Maeda, H. and Konno, T. (1997) Metamorphosis of neocarzinostatin to SMANCS −Chemistry, pharmacology and clinical effect of the first prototype anticancer polymer therapeutic− In *Neocarzinostatin: The Past, Present, and Future of an Anticancer Drug.* (eds. H. Maeda, K. Edo and N. Ishida) pp. 227-267, Springer-Verlag, Tokyo.
[120] Maeda, H., Akaike, T., Sakata, Y. and Maruo, K. (1993) Role of bradykinin in microbial infection: Enhancement of septicemia by microbial proteinases and kinin. In *Agents and Actions Supplements* 42, Proteases, Proteases Inhibitor and Protease-Derived Peptides: Importance in Human Pathophysiology and Therapeutics. (eds. J.C.Chironis and J.E. Repine) pp.159-165, Birkhauser Verlag AG, Switz.
[121] Maeda, H., Edo, K. and Ishida, N. eds. (1997) Neocarzinostatin: The Past, Present, and Future of an Anticancer Drug. Springer-Verlag, Tokyo, pp. 1-289
[122] Fojo, T., and Grady, C. (2009) How much is life worth: Cetuximab, non-small cell lung cancer, and the $440 billion question. *J. Natl. Cancer Inst. 101*, 1044-1048.
[123] Hind, D., Pilgrim, H., and Ward, S. (2007) Questions about adjuvant trastuzumab still remain. *Lancet 369*, 3-5.
[124] Editorial (2008) Welcome clinical leadership at NICE. *Lancet 372*, 601.
[125] Sjöblom, T., Jones, S., Wood, L. D. et al (2006) The consensus coding sequences of human breast and colorectal cancers. *Science 314*, 268-274.
[126] Shah, S. P., Morin, R. D., Khattra, J., et al and Aparicio, S. (2009) Mutational evolution in a lobular breast tumour profiled at single nucleotide resolution. *Nature 461*, 809-813
[127] Maeda, H. (2011) Tumor-selective delivery of macromolecular drugs via the EPR effect: Background and future prospects, *Bioconj. Chem. 21*, 797-802.

INDEX

A

aclarubicin, 47
activation of influenza virus, 21
activation of iNOS, 35
activation of kallikrein-kinin cascade, 21
Akaike, Takaaki, 21,57
albumin, 13,20,21,30, 31, 46
albumin-binding properties, 13
amino acid sequence, 8, 52
angina pectoris, 45
angiotensin II (AT), 2, 31, 37, 40, 41, 42
angiotensin II (AT) induced high blood pressure, 41,42
antigenicity, 13
antioxidant activity, 27
antitumor effect of NCS, 11
antitumor protein, 5
ARCO, 10
Asakawa Award, 22
AT-II, 31, 40, 41
AT-II-induced hypertension, 31
AUC (area under the concentration-time curve), 33, 51
augment the EPR, 53
augmentation of the EPR, 41
augumentation of EPR effect, 46

B

bacterial proteases, 1, 19, 21, 22
Bharate, G., 49
bilirubin, 47, 48
biliverdin, 47
bradykinin, 1, 20, 21, 29, 35, 36
bradykinin generation, 21

C

cancer, 2, 18, 26, 39, 42, 43, 46, 47
cancer chemotherapy, 46, 47
cancer treatment, 14
canolol, 27
carbon monoxide (CO), 36, 47
carcinogenesis, 1, 26, 27
CCRF, 7, 8
Chemical and Engineering News, 10, 52
chemical modification, 3, 9, 23, 29
chemical structure of SMANCS, 11
Children's Cancer Research Foundation (CCRF), 7
Children's Hospital Boston, 7, 34
Christophi, C., 35,57
chronic myelogenous leukemia cells, 49

cisplatin, 48
citation frequency, 37
^{14}C-labeled Lipiodol, 16, 17
cleavage of hemagglutinin, 21
clinical development of NCS, 11
clinical effects, 52
CO (carbon monoxide), 36, 48
collagenase (MMPs), 36
Commemorative Gold Medal Award, 22
compartment models, 12, 65
computed tomography (CT), 15
Controlled Release Society, 38
CT images, 16
CT scan images, 31, 41
cytochrome-P450 reductase, 24

D

Dana-Farber Cancer Institute, 7
Daruwalla, J., 34, 57
Davis, California, 3, 30
degradation of DNA, 11
Department of Surgery, 13, 15, 37
details of the EPR Effect, 29
difficult-to-treat cases, 41
discovery of the EPR, 2, 13, 29, 32, 43
disintegration, 34, 49
DIVEMA (divinyl ether and maleic anhydride copolymer), 23, 32
DNA synthesis, 11
doxorubicin, 47
drawbacks of NCS, 13
Duncan, Ruth, 10, 32, 33, 57
Dvorak, H., 36

E

E. K. Frey–E. Werle Foundation, 22
effect of SOD, 25
elevating blood pressure, 40
encapsulated drugs, 49
endocytotic intracellular uptake, 49
endogenous free, 1, 22, 23, 55

endogenous oxidants, 48
endyene chromophore, 52
enhanced permeability and retention (EPR) effect, 12, 30
enhanced vascular permeability, 29
enhancement of EPR effect, 39, 46
EPR effect, 2, 12, 13, 14, 29, 30, 31, 32, 33, 35, 36, 37, 39, 40, 41, 43, 45, 46, 53
EPR effect and molecular weight, 33
EPR effect, augmentation of, 41, 46
EPR effect-based drugs, 53
EPR effect citation, 37
EPR effect, enhancement of, 40, 39
Evans blue, 20, 21, 29, 31
Evans blue-albumin, 21, 31, 46
extravasation, 20, 21, 29, 30, 31, 35, 36

F

Factor XII, 20
factors facilitating the EPR effect, 35
Fang, Jun, 36, 45, 47, 48, 57
Farber, Sidney, 7, 8
Feeney, E. Robert, 3, 5, 51
femoral artery, 14, 15
fluorescein isothiocyanate (FITC), 6
Folkman, Judah, 34
free radicals, 1, 22, 23, 53, 57
Fritz, Hans, 22, 57
Fulbright, 3

G

gallbladder cancer, 41, 42
Gandy, B. Judith, 57
genetic divergence, 53
Glaser, Charles, 8
Greish, Khaled, 40, 47, 48, 57

H

Hageman factor, 20

Hakuaikai hospital, 40
Harvard Medical School, 7, 34, 40
heat shock protein, 36, 47
Helicobacter pylori, 27
hemagglutinin of influenza virus, 21
heme oxygenase-1 (HO-1), 47
hepatic artery, 14, 16, 17, 31, 42
hepatocellular carcinoma (HCC), 16, 39, 42, 52
hepatoma, 15, 16, 31, 32, 52
heterogeneity of EPR effect, 39, 45
Hinuma, Yorio, 8
Hirano,Takashi, 23
HO-1, 36, 47, 49
HO-1 inhibitor, 36, 49
honorary citizen, 32
honorary mayor,32
HPMA (hydroxypropyl methacrylate copolymer), 32
HPMA-copolymer, 33
Hsp-32, 38
hydrophobic, 10, 13
hypertensive state, 40, 41
hypovascular tumors, 39
hypoxia-inducible factor (HIF-1α), 39, 40, 46

I

immunogenicity, 13, 51
immunoglobulin, 30
immunosuppression, 13
inactivation, 51, 52
inducible form of NOS, 35
inflammation, 1, 19, 26, 27, 35, 36, 53
influenza virus, 1, 21, 22, 23
influenza virus infection,1, 21, 22, 23, 25, 27
iNOS, 26, 35
Institute of Macromolecular Chemistry, 32
intracellular uptake, 9, 49
involvement of proteases, 22
ISDN; isosorbitol dinitrate, 46

Ishida, Nakao, 5, 7, 11, 52
Iwai, Ken, 17
Iyer, A. K., 48

J

Japan NO Meeting, 55
Japan Society of Drug Delivery System, 38, 55, 59
Japanese Society for Bacteriology, 22, 55
Journal of Drug Targeting, 43, 55
Journal of Medicinal Chemistry, 10

K

kallikrein-kinin cascade, 1, 20, 21, 22, 35
Kamata, Ryuji, 19
Kanamaru, Ryunosuke, 11, 57
Kawauchi, Hiroshi, 6
Kimura, Masami, 35, 57
kinin, 1, 20, 21, 22, 29, 35
kinin generation, 22, 29
Kobe University, 55
Koch, Robert, postulate of, 1, 23
Konno, Toshimitsu, 13, 14, 15, 16, 17, 32, 57
Kumagai, Katsuo, 6, 11
Kumamoto, 8, 9, 32, 40, 52, 55, 57
Kumamoto University, 9, 55
Kumamoto University Hospital, 13
Kumamoto University Medical School, 8, 9, 21
Kuraray Company, 10
Kuromizu, Kenji, 8
Kuwahara, Hideo, 27

L

laparotomy, 14, 15
Li, Chang, 40
Li, Chao Hao, 8
Linus Pauling, 23

Lipiodol, 14, 15, 16, 17, 52
liquid ammonia, 8
low vascular density, 39
low vascular density tumor, 40
lung cancer, 46
lymphatic clearance, 17, 37
lymphatic metastasis, 9
lymphotropic drug, 9

M

macromolecules, 2, 9, 17, 30, 37, 40, 46
macrophages, 13
Maki, Shojiro, 32, 57
Maruo, Keishi, 21, 57
massive metastatic liver cancer, 43
massive renal cell carcinoma, 42
Matsumoto, Koki, 19, 20, 57
Matsumura, Yasuhiro, 29, 30, 31, 35, 57
Meares, Claude F., 30
mechanism of mutation, 27
metastatic liver cancers, 39, 42, 43
micelle formation, 47
micronodules, 34
minimum inhibitory concentration, 11
molecular target drugs, 52, 53
molecular weight and AUC, 33
Molla, Akhteruzzaman, 20, 57
Moncada, Salvador, 55, 57
Mongolian gerbil, 27
Müller-Esterl, Werner, 22
mutant virus, 26
mutants of virus and bacteria, 24
mutation, 26, 27, 53
myocardial infarction, 45

N

Nagai Award, 38, 59
Nagai Innovation Award for Outstanding Achievement, 38
Nagamitsu, Akinori, 40
Nakamura, Hideaki, 49

National Cancer Center Hospital East, 29
National Cancer Institute, 5
National Institutes of Health (NIH), 6
natural killer (NK) cells, 13
NCS, 1, 5, 6, 8, 9, 10, 11, 12, 13, 23, 30
Neocarzinostatin (NCS), 1, 5, 52
nitration of G, 24
nitric oxide, 2, 23, 36, 55
nitric oxide synthase, 23
nitrite ion (NO_2^-), 45
nitroglycerin, 2, 45, 46
NO, 23, 24, 26, 47, 48
NO releasing agents, 47
NOS, 24, 26

O

O'Brien, Paul, 34
Oda, Tatsuya, 23
Okamura, Ryoichi, 19
oncogene *bcl/abl,* 49
one-compartment model, 12,
Ono, Yasushi, 11
ONOO⁻, 24, 26, 36
ophthalmology, 19
Osaka University, 55
ovalbumin, 30
ovarian cancer, 44
ovomucoid, 3, 30

P

pancreatic cancer, 42, 43, 44
pegylated proteins, 14, 52
PEGylated ZnPP, 47
PEG-ZnPP, 47, 48
peroxynitrite, 24, 26, 36
pharmacokinetics, 9, 11, 12, 32, 48, 51
pharmacokinetics of NCS, 11, 12
pin point delivery, 34
pinpoint targeting, 2
pirarubicin, 34, 47, 48
plasma concentration, 51

pO_2, 45, 46
poly(styrene-co-maleic anhydride) (SMA), 10
polymer leakage, 34
polymer-conjugated superoxide dismutase (SOD), 22
Postulates of Robert Koch, 1, 23
protease, 1, 3, 19, 21, 22, 29, 51, 55
protein drugs, 51, 52, 53
Pseudomonas aeruginosa, 19
pyran copolymer, 23
pyran-conjugated SOD, 23, 25

R

Raeder, Beate, 49
reactive nitrogen species (RNS), 1, 19
reactive oxygen species (ROS), 1, 19, 48, 52
Regelson, William, 32
renal cell carcinoma, 39, 42, 44
renal excretion, 12
Ringsdorf, Helmut, 32, 38, 43, 57
ROS, 1, 19, 23, 26, 27, 48
ROS and RNS in infection, 26
Royal Pharmaceutical Society, 43

S

Sahoo, S. K., 49
Salmonella, 27
San Antonio, 32
scanning electron micrographs of vascular casts, 34
scavenger of NO, 35
Seki, Takahiro, 45
Seldinger's method, 15
Sendai virus, 5, 26
Serratia marcescens, 19
serratial 56K protease, 21
Seymour, Len, 32, 57
SMA, 1, 10, 13, 30, 34, 47, 48, 49, 51
SMA micelles, 47, 48, 49

SMANCS, 1, 2, 10, 13, 16, 30, 31, 32, 40, 41, 43, 44, 46, 51, 52,
SMANCS application, 16
SMANCS/Lipiodol, 14, 15, 16, 17, 31, 40, 46
SMANCS/Lipiodol infusion, 46
SMANCS/Lipiodol injection, 31
SMANCS, structure of, 11
SMANCS synthesis, 10, 37
SMA-pirarubicin micelles, 34, 50, 70
SMA-THP (pirarubicin), 49
SMA-ZnPP, 47, 49
Society of Cancer Prevention of Japan, 55
SOD, 1, 22, 23, 25
solid tumor tissues, 30
solid tumors, 2, 20, 30, 32, 35, 37, 39, 53
State of Oklahoma, 32
stomach cancer, 42, 43
structure-activity relationships, 9
super oxide, 1, 22, 23, 25
super oxide dismutase [SOD], 1, 22, 23
Suzuki, Maro, 40

T

Takaki, Asami, 57
Takeshita, Jiro, 10
Taniguchi, Naoyuki, 55
Tatematsu, Shoei, 27
taxol, 47
the B-type staining (peripheral ring shape), 43
therapeutic effect of SMANCS (SX)/Lipiodol (LP), 42
THP, 47, 48, 49
time dependent increase of tumor uptake, 33
tissue distribution, 9, 51
tissue oxygen tension (pO_2), 45, 46
Toda, Noboru, 57
Tohoku University, 3, 5, 11
Tohoku University Medical School, 5, 11

transferrin, 30
tumor necrosis factor (TNF)-α, 36
tumor nodules, 34
tumor uptake of macromolecules, 30
tumor-selective drug delivery, 2, 16, 34
two-compartment model, 12

vesicorenal recirculation, 12
viral infection, 1, 22, 23, 49
virus infectivity, 21, 22
vivo half-life, 23, 51
VPF/VEGF, 36
VX-2, 14, 16

U

ubiquitous characteristics, 53
Ulbrich, Karel, 32
University of California, 3, 8, 30
University of Melbourne, 34
University of Vienna, 49
urinary bladder, 12
urinary excretion, 12, 51, 52

V

Valent, Peter, 59
vascular density, 37, 39, 40
vascular endothelial cell growth factor (VEGF), 36
vascular extravasation, 20
VEGF, 36, 40, 46

W

Walker 256 tumor, 6

Y

Yamanouchi Pharmaceutical Company, 11, 52
Yokoyama, Ikuzo, 15, 16
Yokoyama, Mitsuhiro, 55

Z

zinc protoporphyrin (ZnPP), 50, 71
ZnPP, 47, 48, 49